SARK IN THE DARK

Sark in the Dark:
Wellbeing and Community on the Dark Sky Island of Sark

SOPHIA CENTRE MASTER MONOGRAPHS: VOLUME 3

Sophia Centre for the Study of Cosmology in Culture
University of Wales Trinity Saint David
Jennifer Zahrt, General Editor

Sark in the Dark

WELLBEING AND COMMUNITY ON THE DARK SKY ISLAND OF SARK

by Ada Blair

SOPHIA CENTRE PRESS

First published by Sophia Centre Press in 2016.

SOPHIA CENTRE PRESS
University of Wales Trinity Saint David
Ceredigion, Wales SA48 7ED, United Kingdom
www.sophiacentrepress.com

ISBN 978-1-907767-42-5
British Library Cataloguing in Publication Data.
A catalogue card for this book is available from the British Library.

Book Design by Joseph Uccello.

Printed in the UK by LightningSource.

PERSONAL INTRODUCTION

This monograph originally began life in 2014 as part of my research for the MA in Cultural Astronomy and Astrology at the University of Wales Trinity Saint David. In the course of my studies, I undertook a research project exploring the role that the night sky plays in relation to the Dark Sky island community of Sark using the qualitative method of intuitive inquiry. I wanted to investigate whether a community living in close relationship with its dark sky might experience similar beneficial and transformative effects to those that have been reported in relation to 'green'/ grounded nature.

In the first chapter I introduce the island of Sark and its wonderful dark sky, probing the nature of islandness and documenting Sark's journey to becoming a Dark Sky Community. Then the next chapter explores the human desire to see the night sky, the commercialisation of this desire through tourism, heritage and nostalgia. I examine fear of the dark, the 'nature' of nature, and nature and wellbeing. Then I spend a chapter documenting my research, featuring direct interviews with residents of Sark, and another chapter revealing my findings. Finally, in the closing chapter I raise present and future perspectives and some areas that might warrant further inquiry into Dark Sky culture and what we can learn from the Sark island community.

To my parents,
Annie Smith Blair and Thomas Deans Blair,
who struck the spark.

ACKNOWLEDGEMENTS

Without the patience, goodwill, love and help from a number of people this book would not have seen the light of day. A huge thanks to the people of Sark who generously gave of their time, my friend Alex White who first introduced me to Sark and set everything in motion, Dr Bernadette Brady, who supervised my dissertation on the MA in Cultural Astronomy and Astrology, Dr Jenn Zahrt my editor at the Sophia Centre Press who encouraged me to use my own voice, Lydia Bourne and Sue Daly who gifted some of the photographs, Graham and Sheila Harker for accommodation and as always Chris Barry my partner and fellow traveller.

TABLE OF CONTENTS

Chapter 4

Chapter 5

FOREWORD

by Marek Kukula

IN 1675 KING Charles II gave orders for an astronomical observatory to be built on a hill beside his palace in Greenwich. This Royal Observatory was intended to provide a more suitable working environment for Charles' newly appointed Astronomer Royal, John Flamsteed, who for several months had been conducting his studies of the night sky from the roof of the Tower of London. Legend has it that the move to Greenwich was down to the Tower's famous ravens whose droppings were interfering with the Astronomer Royal's telescopes, although a more likely source of annoyance for Flamsteed might have been the constant smoke from thousands of London chimneys. But, whatever the real reason for the move, it certainly wasn't a lack of darkness: in the seventeenth century, long before the advent of gas and electric lighting, London's skies would have been pitch black.

Fast-forward three centuries and the picture is very different. London has engulfed the village of Greenwich and at night the sky above the city is awash with the scattered glow from buildings, street lamps and security lights. From the city centre only a few dozen of the brightest stars are visible to the naked eye and the glittering band of the Milky Way is completely hidden from view. This situation is replicated in urban areas around the planet and, with city dwellers making up more than half the world's population, this means that billions of people now never experience the sight of a truly dark sky full of stars. It's hard to imagine how a modern city could function without artificial light but, in embracing the convenience of twenty-four hour illumination, we have, perhaps unwittingly, let go of our age-old connection to an important part of our natural heritage: the sky.

There is however a shadow of hope among all this blinding illumination. Where there is a will, light pollution can be minimised: sensible lighting design can ensure that light in our towns and cities goes down onto the streets where it's needed and not up into the sky where it simply blocks our view of the stars, while in rural areas the same approach can help to protect what darkness remains, saving both energy and money in the process. More and more people are coming to value the sky as a part of their environment that's worth preserving and organisations such as the International Dark Sky Association have played an important role in both raising awareness of the issue and helping communities to take active steps to cherish and protect the natural beauty above their heads. But what, apart from an inspiring view, are the benefits of a starry sky? Already studies have been carried out on the possible effects of light pollution on wildlife and even on human health but in this volume Ada Blair explores another, equally important dimension of the dark sky debate. For our relationship with the environment is not just physical, but emotional and imaginative too, and by focusing on the Dark Sky Island of Sark, its people and its stories, we begin to see how individuals and communities might be affected by the presence of cosmic immensities in their everyday lives.

Indeed, perhaps tiny Sark has a lesson for us all. The island's Dark Sky status is the result of a community of people coming together to protect their common heritage. Its splendid night sky should serve as a reminder that the Earth itself is an island, and one that we must all learn to value and share together.

Dr Marek Kukula
PUBLIC ASTRONOMER,
Royal Observatory Greenwich,
November 2016

PREFACE:

WELLBEING AND DARK SKIES

by Nicholas Campion

IT IS A pleasure to publish this book by Ada Blair, documenting her findings on the benefits of living with dark night skies. It is difficult to establish whether any work has been done in this area until now. Even if it has, Blair's is amongst the first. Indeed, if one googles, for example, 'benefits of living under dark skies', the only search result in English is Blair's blog post on the topic. And as Blair herself said,

> Although dark sky supporters often claim dark skies enhance wellbeing, there is a little research to support these claims. Much of the research focuses instead on the nega-

> tive impacts of light pollution on human and animal health and behaviour.[1]

Her assertion is correct. For example, the official website promoting the Brecon Beacons International Dark Sky reserve in south Wales reports that,

> New research has revealed that light pollution not only limits the visibility of stars, but also disturbs the navigational patterns of nocturnal animals. This has contributed to the decline of many of our native nocturnal species.[2]

The dark sky movement, catalysed by the International Dark-Sky Association (IDA), aims to reduce the modern curse of light pollution in order to restore the sight of the rich, beautiful, awesome starry sky, which has been largely lost to most people only over the last fifty years. The movement is big, and getting bigger, even if it hasn't made a breakthrough to the forefront of cultural consciousness. It faces stiff, passive resistance from the status quo. For example, arguments about the costs of light pollution tend to have little clout as compared to concerns over safety. So you can see the stars, the sceptics say, but what use is that if you get run over or mugged? This is why Blair's book,

1 Ada Blair, *The Psychology of Dark Skies*, International Dark-Sky Association, 27 July 2016, http://darksky.org/the-psychology-of-dark-skies/ http://www.darksky.org/about-ida [Accessed 24 Oct 2016].

2 'About the Brecon Beacons International Dark Sky Reserve', Brecon Beacons: Our National Park, http://www.breconbeacons.org/about-brecon-beacons-dark-sky-reserve [Accessed 24 Oct 2016].

documenting her research at the University of Wales Trinity Saint David, is important. Throughout her investigation Blair found that experience of the night sky is beneficial for individual wellbeing. This conclusion comes up again and again in her findings. Wellbeing is now a subject of government concern, and substantial reports have been produced, for example, by the UK's Department for Environment, Food and Rural Affairs (DEFRA).[3] There are even official statistical studies which rely on the following standard definition of wellbeing, which I will quote in full:

> Wellbeing is a positive, social and mental state; it is not just the absence of pain, discomfort and incapacity. It arises not only from the action of individuals, but from a host of collective goods and relationships with other people. It requires that basic needs are met, that individuals have a sense of purpose, and that they feel able to achieve important personal goals and participate in society. It is enhanced by conditions that include supportive personal relationships, involvement in empowered communities, good health, financial security, rewarding employment and a healthy and attractive environment.[4]

3 Gemma Harper and Richard Price, *A Framework for Understanding the Social Impacts of Policy and Their Effects on Wellbeing*, A paper for the Social Impacts Taskforce Defra Evidence and Analysis Series Paper 3 (London: Department for Environment, Food and Rural Affairs (DEFRA) April 2011.

4 Louise Skilton, *Working Paper: Measuring Societal Wellbeing in the UK* (London: Equalities and Wellbeing Branch Office for National Statistics May 2009), p. 6.

Wellbeing, the report states, is a comprehensive condition resulting from the state of one's total relationship with the world: it is central to living in an empowered community. For DEFRA wellbeing concerns not just individuals but the entire UK: a country can have wellbeing, or suffer from a lack of it. The Welsh government has even incorporated wellbeing principles into its policy making, in the 'Well-Being of Future Generations (Wales) Act 2015'. The Act has seven core goals, of which I will select the following: under 'A resilient Wales', it is stated that the country should maintain and enhance 'a biodiverse natural environment with healthy functioning ecosystems that support social, economic and ecological resilience'; the section on 'A healthier Wales' calls for a society 'in which people's physical and mental well-being is maximised and in which choices and behaviours that benefit future health are understood'; and under 'A Wales of vibrant culture', we read that the country should promote and protect 'culture (and) heritage'.[5] All these aims have been incorporated in the sustainability agenda of the University of Wales Trinity Saint David, where Blair studied, via the Institute of Sustainable Practice, Innovation and Resource Effectiveness (INSPIRE). What do I apply to dark skies from the Act's goals? I think there are three fundamental points: that the sky is (1) an

5 Llywodraeth Cymru/ Welsh Government, *Well-Being of Future Generations (Wales) Act 2015: The Essentials*, p. 6, http://www.cynnalcymru.com/wp-content/uploads/2016/08/Guide-to-the-WFGAct.pdf http://www.darksky.org/about-ida [accessed 24 October 2016]; Llywodraeth Cymru/ Welsh Government, *Well-Being of Future Generations (Wales) Act 2015*, http://www.legislation.gov.uk/anaw/2015/2/pdfs/anaw_20150002_en.pdf

integral part of our diverse natural environment, and essential to a resilient ecosystem; (2) central to wellbeing and (3) vital to a full understanding of cultural heritage, if we are to properly understand past generations' and other cultures' use of the sky. But here I will focus on wellbeing.

It is now well known that contact with natural environments benefits wellbeing. Communication with animals and birds, walking through forests, swimming in rivers, sitting on mountain tops feeling the wind and gazing at a view are all benevolent and benign. A recent literature review compiled in the UK by The Wildlife Trusts for the University of Essex contained this conclusion:

> Overall there is a large body of evidence from published peer-reviewed and grey literature to suggest that contact with a wide range of natural environments can provide multiple benefits for health and wellbeing.
>
> These benefits from nature include improvements to physical health (through increased physical activity); and improvements to psychological and social wellbeing, in a number of ways, including: reductions in stress and anxiety, increased positive mood, self-esteem and resilience, improvements in social functioning and in social inclusion.[6]

6 Rachel Bragg, Carly Wood, Jo Barton and Jules Pretty, *Wellbeing Benefits from Natural Environments Rich in Wildlife: A Literature Review of The Wildlife Trusts*, (Newark: The Wildlife Trusts and Colchester: University of Essex, 2015), https://www.wildlifetrusts.org/sites/default/files/wellbeing-benefits-fr-nat-env-report-290915-final-lo.pdf, p. 5 [Accessed 24 Oct 2016].

There seems to be little that contact with nature cannot benefit. But, the survey continues, more can be done

> Increasing access to a wide range of nature-based activities within society will provide benefits to public health and provide savings to the UK economy.[7]

But what more? My answer is that the sky should be included part of the natural environment. After all, it is actually at least half of the natural environment, measured by the visual area it covers. It is also pretty important. Without it we would die. There would be no sun, no heat, no light, and no fresh water from rain. We would exist in a frozen, pitch black, barren world. Or, actually, we would not exist at all. One target is therefore to draw the attention of environmental and sustainability campaigners to the sky's overwhelming importance. It reached environmental consciousness once, in the 1980s, when we became aware of the hole in the ozone layer caused by CFCs. But there is more. As Blair wrote,

> The fields of ecopsychology and environmental psychology look at how encounters with nature may be beneficial and transformative but usually the focus is on 'green'/grounded nature rather than encounters with the sky. In fact, it's not often that the sky is even considered part of nature.[8]

7 Bragg, et al., *Wellbeing Benefits*, p. 5.
8 Blair, 'Psychology'.

The IDA has had some success. As its website announces

> Our award-winning flagship conservation program recognizes and promotes excellent stewardship of the night sky. We've certified more than 65 Dark Sky Places worldwide across six continents, comprising more than 58,000 square km (21,200 square miles).[9]

Written testimony concerning the value of the night sky is found throughout the world. From the tradition inherited by the modern west, let me cite just a few. From the Jewish scriptures we find this familiar, inspirational refrain: 'The heavens recite the glory of God, and the sky tells of the work of His hands'.[10] In the classical world the first century CE Alexandrian polymath Claudius Ptolemy, perhaps the most important influence on medieval and Renaissance astronomy, followed with equally awe-struck. He declared,

> Mortal as I am, I know that I am born for a day, but when I follow the serried multitude of the stars in their circular course, my feet no longer touch the earth; I ascend to Zeus himself to feast me on ambrosia, the food of the gods.[11]

9 'Our Work', International Dark Sky Association, http://darksky.org/our-work/; for a map of dark sky places see http://darksky.org/idsp/finder/ [Both Accessed 25 Oct 2016]

10 *The Complete Jewish Bible with Rashi Commentary*, http://www.chabad.org/library/bible_cdo/aid/16240/jewish/Chapter-19.htm [Accessed 7 Apr 2016].

11 Claudius Ptolemy, 'Anthologia Palatina', ix.577, quoted in Franz Cumont, *Astrology and Religion Among the Greeks and Romans* (New York: Dover, 1960), p. 81.

Ptolemy's contemporary, the Roman emperor Marcus Aurelius, shared his enthusiasm: 'Survey the circling stars', the emperor wrote, 'as though you yourself were in mid-course with them. Often picture the changing and rechanging dance of the elements. Visions of this kind purge away the dross of our earth-bound life'.[12] The soul, Marcus Aurelius believed, could be cleansed by an imaginative union with the stars. Of course it could, for the sky was beautiful. Coming to the beginning of the modern world, the philosopher Emmanuel Kant write in 1788 that,

> Two things fill the mind with ever new and increasing admiration and awe, the oftener and the more steadily we reflect on them: *the starry heavens above and the moral law within.* I have not to search for them as though they were veiled in darkness or were in the transcendent reason beyond my horizon; I see them before me and connect them directly with the consciousness of my existence.[13]

The stars are exquisite, beautiful and inspiring in themselves, but as Kant suggested there is a direct connection between them and us. We are part of the same cosmos. We are absolutely interconnected and interdependent. Moving forward to the twentieth century, here is Fred Hoyle,

12 Marcus Aurelius, *Meditations*, trans. Maxwell Staniforth (Harmondsworth, Middlesex: Penguin, 1964), V.47, p. 112; see also IX.29, p. 144. See Plato, *Republic*, 2 Vols, trans. Paul Shorey (Cambridge, MA: Harvard University Press, 1937), 516B.

13 Immanuel Kant, *Critique of Practical Reason*, Great Books of the Western World 42, (London: Encyclopaedia Britannica, 1952), pp. 360–61.

perhaps the most famous British astronomer of the 1960s and 70s:

> Our everyday experience even down to the smallest details seems to be so closely integrated to the grand-scale features of the Universe that it is well-nigh impossible to contemplate the two being separated.[14]

And this is from a report on a 1988 planetarium show by British astronomer Heather Couper:

> And our connections with the stars go deeper than our ancestors ever imagined. Dead heroes do not in fact make up the star-patterns in the sky: but the matter of the stars does make our planet, and our bodies. 'And we ourselves', Couper concludes, 'are made of star dust'.[15]

Carl Sagan, one of the USA's greatest science popularisers of the last century, is the author of some of the most inspiring statements on humanity's relationship with the cosmos. He also decried our modern distance from it.

> We have grown distant from the Cosmos. It has seemed remote and irrelevant to everyday concerns, but science has found not only that the universe has a reeling and ecstatic

14 Fred Hoyle, *Frontiers of Astronomy* (Charleston, SC: BiblioBazaar, 2011), p. 304.

15 Marcus Chown, 'The Life and Death of a Star', review of 'Starburst! New Show by Heather Couper at the London Planetarium', *New Scientist*, 21 April 1988, p. 63.

> grandeur, not only that it is accessible to human understanding, but also that we are, in a very real and profound sense, a part of that Cosmos, born from it, our fate deeply connected with it. The most basic human events and the most trivial trace back to the universe and its origins.[16]

But what has made us distant from the cosmos? What separates us from our origins? Light pollution.

We can make the inspirational case for dark skies, but for the purposes of dealing with planners and policy makers, we can also make pragmatic arguments. For example, dark skies can attract tourists. To stay with our Welsh example, we can point to the Brecon Beacons International Dark Sky Reserve, but there are examples all over the world.[17] In other research, also in the UK, at Nottingham Trent University, Daniel Brown found that exposure to dark skies encourages us to reduce light pollution and so save energy.[18] And if, as Blair shows, increasing wellbeing is of economic benefit, and dark skies can increase wellbeing, then bringing back

16 Carl Sagan, *Cosmos: the Story of Cosmic Evolution, Science and Civilisation* (London: Warner Books, 1994), p. 12.

17 '5 top spots for stargazing around Brecon', http://www.visitwales.com/explore/mid-wales/brecon-beacons/stargazing [Accessed 24 Oct 2016]; 'About the Brecon Beacons International Dark Sky Reserve', Brecon Beacons: Our National Park, http://www.breconbeacons.org/about-brecon-beacons-dark-sky-reserve, [Accessed 24 Oct 2016]; Brecon Beacons National Park Authority, 'Brecon Beacons Dark Skies time-lapse', https://www.youtube.com/watch?v=YG--TieOrvo.

18 Daniel Brown, *How can Higher Education Support Education for Sustainable Development? What can Critical Place-Based Learning Offer?* [Unpublished MS, University of Nottingham, 2013].

dark skies, restricting unnecessary street lighting is of economic benefit.

Light pollution cuts off our heritage, reduces our wellbeing and deprives us of contact with a huge part of our natural environment. Imagine if we woke one day and were unable to see the green fields and hills of Wales or, for that matter, the forests of the Amazon, the mountains of Nepal or the great rivers of the world. But that is what we have done, and are doing, with the sky, impoverishing our lives in the process. Hopefully Blair's book will play a small role in the wider effort to slow down, or even reverse, that process.

Nicholas Campion,
SOPHIA CENTRE PRESS,
Sophia Centre for the Study of Cosmology in Culture,
Faculty of Humanities and the Performing Arts,
University of Wales Trinity Saint David.

CHAPTER 1:

SARK THE ISLAND AND ITS DARK SKY

SARK IS A beautiful unspoilt island with forty miles of craggy coastline providing a sanctuary for thousands of species of plants, birds, insects and mammals (including the black rat, one of Britain's rarest mammals) and peaceful leafy lanes with high hedgerows. At only five and a half square kilometres, it is the smallest of the four main Channel Islands, which lie between England and France. In fact Sark is actually two islands, Big Sark and Little Sark, which are joined by the narrow isthmus La Coupee, which has precipitous 260 feet drops on either side to Grande Greve beach below. Before there were railings it was not unknown on windy days to have to crawl across on hands and knees to avoid being blown off into the sea. The island has a resident population of about 550 and attracts up to

40,000 visitors annually, predominantly between May to September. Its economy is heavily dependent on tourism and financial services. The Avenue is the island's 'village' centre and its one shopping area. It is Sark's 'high street' with an eclectic range of small shops; at Christmas time it is the one area of Sark where public Christmas lights are permitted. Sark has no National Health Service, (residents and visitors have to pay for any medical care themselves), no cash points, no chain stores, no public street lighting (locals use head torches or the stars to navigate at night) and no paved roads or pavements.

All cars and vehicles other than tractors, 'invalid carriages', construction vehicles, or combine harvesters are banned, and people travel by bike or on foot. Rush hour does not exist here! An islander has to demonstrate that their business or occupation requires a tractor before a licence will be granted. Tractors must be off the roads between 10 PM and 6 AM, and none are permitted on Sundays other than in the case of an emergency or for stock feeding. Although there are a small number of horse-drawn carriages, these are mainly used by visitors or by locals for special occasions. The island has no airport, no airstrip, and most people travel to Sark by an Isle of Sark Shipping Company ferry from the larger island of Guernsey, a journey of just under an hour.

On disembarking from the ferry at Maseline Harbour the 'toast rack', a tractor-pulled carrier, takes most passengers up the steep road through a wooded valley to the top of Harbour Hill just next to the conveniently located historic Bel Air Inn. A tractor from Jimmy's Carting Services will then transport visitors' luggage to where they are staying but is not permitted to carry passengers, so onward passage involves a walk or cycle ride. Sark Ambulance Service's two ambulances and Sark Fire and Rescue Service's three appliances are all also pulled by tractor and the services staffed by volunteers. With its annual sheep racing weekend in July, weekly meat raffle on Friday nights at the Bel Air pub and scarecrow competitions, it is an island whose way of life is often, perhaps unsurprisingly, described as 'idiosyncratic'.

As to myself, I am a psychotherapist who, despite always having lived in Scottish towns and cities, has a longstanding interest in how encounters with nature (including the sky) may have a positive impact on our wellbeing – perhaps even leading to some kind of personal transformation or 'eureka' moment. I have experienced a number of such moments myself, most notably ten years ago when I spent five days on Knoydart, on the west coast of Scotland, a place often referred to as Britain's last wilderness. Incidentally, like Sark, Knoydart is usually reached by boat. Since the late 1990s I have studied ecopsychology, a subject which, in the words of one of its founders Theodore Roszak, aims to, 'bridge our culture's long-standing historical gap between the ecological and the psychological' and to give some insights into what we mean by 'nature' and the type of relationship humans have with it.[1] Ecopsychology proceeds from the assumption that at a deep level our psyche is in a

1 Theodore Roszak, *The Voice of the Earth* (New York: Simon & Schuster, 1992), p. 14.

Midnight, The Avenue © www.bournephotography.me.uk

close relationship to the earth and that to begin to realise this connection, rather than seeing nature merely as a backdrop to our lives, is considered to be healing for both.[2] And after all us humans are not only made of stardust, but are animals ourselves.

For over three decades now I have lived in Edinburgh, Scotland, a beautiful historic capital city, but unfortunately, like many urban areas worldwide, a city with a lot of light pollution where it is usually only possible to see at most thirty stars even on a clear night. My experiences of seeing lots of stars have usually been restricted to trips visiting friends in rural Scotland or, most spectacularly, whilst on holiday in the desert in northern Africa.

When Alex White, a close friend of mine, started working on Sark in 2009 and began to tell me that things were 'a bit different here' and how much I would love the dark skies, I was intrigued. Hearing about him winning a prize for his lavender shortbread in the Grand Autumn Show and getting the part as one of three Dorothys in a production of 'The Wizard of Oz' had already convinced me his life there was very different from his previous one in Edinburgh! However, in many ways I was 'in the dark' about Sark when I first arrived there with my partner for a few days holiday in 2010. As far as I was concerned, the main purpose of our trip was to visit Alex. I had limited prior knowledge of the island, other than faint recollections of watching the Channel 4 adaptation of Mervyn Peake's novel *Mr. Pye*. Needless to say, I was unprepared for how scared I felt when I tried

2 Roszak, *Voice of the Earth*, p. 5.

to cycle in the dark for the first time and just how dark its night sky is (although there is a small amount of light pollution from neighbouring islands and the east coast of France). However I also enjoyed the fact that when you have no option but to walk or cycle at night someone's wealth or status is not immediately obvious - darkness is a great equaliser! Gradually, when I became more used to being out at night, enveloped by Sark's very dark sky, I regularly noticed a positive effect on my wellbeing and a heightened sense of my insignificance within the larger cosmos. I became curious if others have had similar experiences.

Nicholas Campion states there is no society that, 'does not express at least some fascination with the sky', and I assumed the same would be true particularly of the community of Sark.[3] So when it came to choosing a research topic for my MA dissertation there was no doubt in my mind that I wanted to pursue this idea further. Also it gave me more reasons to visit Sark! Throughout the process of conducting the research another theme began to emerge which I found increasingly hard to ignore. Although Sark has a clear head start in terms of accessing unspoilt night skies and a community committed to preserving this, what difference could people like me, living in more light-polluted areas make to improve our views of the night sky? Also, what might the barriers be to achieving this?

Although Sark is probably best known for its lack of public lighting and cars, it also boasts many unique charac-

3 Nicholas Campion, *Astrology and Cosmology in the World's Religions* (New York: New York University Press, 2013), p. 1.

teristics. It has a voluntary police force and the crime rate is enviably low. Sark Prison is the world's smallest prison, built to hold only two prisoners, and is still occasionally used mainly for cases of drunk and disorderly behaviour. Tractors being used at unsociable hours also account for a sizable percentage of reported crime. Politically, the island is a royal fief (having a feudal connection to the English Crown) in the Bailiwick of Guernsey, a Crown Dependency with its own set of laws based on Norman law, and it has its own parliament. The island became a fief in 1565, which resulted in Hellier de Carteret, Sark's first Seigneur (Lord), arriving with the forty settler families charged with guarding the island against pirates. Before this formal settlement it had been home to a disparate group of people including pirates, monks and French armed forces. In 2008, following the first general election on the island under new constitutional arrangements (when 28 conseillers were elected from a total of 57 candidates), its previous system of government, a legislative body known as the Chief Pleas, was dismantled and reforms were introduced. Until then Sark was considered to be the last feudal state in Europe with the Seigneur holding the island from the Crown in perpetuity, and governing in conjunction with Chief Pleas, which then consisted of unelected tenement holders and twelve elected deputies. The island has largely welcomed the advent of democracy and average turn-out for elections is now around 80% with one conseiller for every fifteen residents. In 2012, following a survey of the population's perceptions of Sark and in an effort to, 'maintain the principles of a society that has served Sark so well for so many years', a document 'A

Vision for Sark' was developed by the Chief Pleas.[4] Its aim is to guide the work of the Committees of the Chief Pleas, setting out how future change could be achieved and recognising that, 'Sark can no longer rely on everyone abiding by largely unwritten rules and "doing the right thing."'[5] However some residents do not feel the reforms go far enough and are seeking a repeal of the 2008/2010 Reform Law. In 2014 they formed a campaigning group, 'Sark First' to push for an alternative system of government. Meanwhile although Sark is not a sovereign state, it continues to be neither part of the United Kingdom nor the European Union, is financially independent of both the United Kingdom and Guernsey, and has no income tax or VAT.

Two churches cater for the island's spiritual needs: St. Peter's which is within the Anglican Deanery of Guernsey (in the Diocese of Winchester) and a Methodist Church which comes under the authority of the Guernsey Methodist Church. Both churches are an integral part of the community and the focus of many community activities. St. Peter's is currently involved in fundraising for much-needed restoration work to be done on the building and the Methodist Church is constructing a Sanctuary Centre that will meet multi-use community and ecumenical needs, provide emergency help to vulnerable people, and support and provide for carers and the elderly.

Sark has a very strong community ethos and the people

4 Sark Chief Pleas, *A Vision for Sark*, www.gov.sark.gg/Downloads/Reports/A_Vision_for_Sark.pdf, [Accessed 16 Feb 2015].

5 *A Vision for Sark*.

place great emphasis on mutual support, as is evidenced by the vast number of community events and fundraising initiatives which take place throughout the year. Jumble sales, quizzes, charity knits, flower and produce shows, cream tea competitions and tapestry weaving are all regular features on the island's calendar. Many events take place at the Island Hall, a community hub, which also has a café and bar. As there is no National Health Service, the cost of most medications for residents is subsidised by the Professor Saint Fund. The Fund originated in a bequest from Professor Charles Saint, a doctor who came to live on Sark after he retired from medicine.

Many projects are entirely funded by community initiatives and Sark School is one such project. The school has four full-time teachers and is run by the Education Committee of the Chief Pleas providing education for around 35 children up to the age of sixteen years, although it is common for children to leave to be educated off the island before sixteen. It broadly follows the UK National Curriculum and the curriculum is adapted to accommodate the needs of Sark children, reflecting the distinct nature of living on the island. Its strong focus on fostering a love of nature is indicated by outdoor education activities such as forest schools, beach schools, Sark Adventurers and Sark Watch (a junior branch of the Wildlife Trust). Lessons take place outside whenever possible as a news item from March 2015 on Sark Tourism's website illustrates,

> Pupils in Sark School's Class Three were all prepared for the eclipse today with a special outdoors science lesson and

> solar viewers. Sadly though the heavy cloud failed to lift and there wasn't even a glimpse of the sun but the light level fell significantly and the children did observe a group of birds returning to some nearby trees to roost thinking it was nightfall.[6]

In his book, *Sark Folklore*, Sark resident Martin Remphry describes *les veilles* or The Watch, the gatherings of family and friends which took place in Sark homes or barns throughout the fifteenth to nineteenth centuries.[7] Mainly held during the dark winter months, these were ostensibly for the purposes of spinning or knitting, but were also ideal opportunities for gossip and the telling of popular tales, such as stories related to witches. Beliefs and superstitions about witches were common throughout the Channel Islands especially during the sixteenth to eighteenth centuries. Judicial records from that time show a number of witchcraft trials took place on nearby Guernsey (with at least four Sark residents being accused), and according to Victor Hugo there was, 'nothing commoner than sorcerers on Guernsey in the 1850s. [...] They exercise their profession in certain parishes, in profound indifference to the enlightenment of the nineteenth century'.[8] There was concern specifically about the possibility of witches flying down unguarded chimneys and taking up permanent

6 Sark Tourism, 'What eclipse?', http://www.sark.co.uk/what-eclipse-9047 [Accessed 13 Dec 2015]

7 Martin Remphry, *Sark Folklore* (Sark: Gateway Publishing Ltd. 2003).

8 Victor Hugo, 'Toilers of the Sea', http://www.online-literature.com/victor_hugo/toilers-of-the-sea/2/ [Accessed 13 Dec 2015].

residence. This is apparently the reason why some of the chimneys on the older houses on Sark have caps on them. As I cycled and walked around the island I also spotted several examples of the witches' seat, stone or step. This is a ledge built on the outside of the chimney originally to stop rainwater getting through the join between the thatched roof and chimney. Its other purpose allegedly however was to tempt witches to rest on it and warm themselves, thus distracting them from trying to come down the chimney.

The absence of public streetlights means Sark is an environment where celestial bodies and sky features are particularly visible at night; the Milky Way is a common wintertime sight. Whereas in most parts of Europe it is unlikely that more than two hundred stars could be seen with the naked eye, even on a clear night, on Sark it would not be unusual to see more than 5,000. I found being able to see huge numbers of stars on Sark quite perplexing at first; it was hard for me to identify constellations I am familiar with such as Cassiopeia, Draco and Orion, difficult to 'get a foothold' in the sky. In fact the 2014 Star Count carried out by volunteers for the Campaign for the Protection for Rural England (CPRE) reckons that only 4% of the UK population can normally see more than thirty stars on a good night.[9] Furthermore, 59% of those who took part in the 2014 Count saw ten stars or fewer within the constellation of Orion – an indication of severe light pollution in their area. In cer-

9 CPRE, 'Star Count 2014: A Dark Outlook for Starry Skies', http://www.cpre.org.uk/media-centre/latest-news-releases/item/3583-star-count-2014-a-dark-outlook-for-starry-skies [Accessed 23 Nov 2015].

tain areas of London it is no longer possible to see Orion at all. In the US the situation is even worse with more than 65% of the population there unable to see the Milky Way with the naked eye.[10]

In January 2011, after an application process spearheaded by committed locals, the US-based non-profit International Dark-Sky Association (IDA), founded in 1988, designated Sark as a Dark Sky Community making it the world's first Dark Sky island. Moreover, as Paul Bogard notes, unlike many Dark Sky Places, 'what makes Sark especially compelling is that people actually live there'.[11] Becoming a Dark Sky Community requires that community to demonstrate 'exceptional dedication to the preservation of the night sky through the implementation and enforcement of quality lighting codes, dark sky education, and citizen support of dark skies'.[12] In a letter supporting Sark's application for Dark Sky status, the President of the Sark Chamber of Commerce, Peter Tonks, notes the Chamber is, 'very much aware of the value of our night sky as an attraction for tourism'.[13] Therefore Tonks was acknowledging Sark's

10 P. Cinzano, F. Falchi and C. D. Elvidge, 'The First World Atlas of Artificial Night Sky Brightness', *Monthly Notices of the Royal Astronomical Society* 328.3 (2001): pp. 689–707, p. 689.

11 Paul Bogard, *The End of Night: Searching for Natural Darkness in an Age of Artificial Light* (London: Fourth Estate, 2013), p. 185.

12 International Dark-Sky Association, 'About IDA', http://www.darksky.org/international-dark-sky-places/about-ids-places/communities [Accessed 17 May 2014].

13 Peter Tonks in Steve Owens, 'Sark Dark Sky Community: A Dark Sky Island', http://www.ida.darksky.org/assets/documents/dark%20sky%20community%20application.pdf [Accessed 2 Mar 2014].

dark sky could further enhance the island's distinct local identity.

THE NATURE OF ISLANDNESS

Sark is an island, and islands – along with forests and seashores – are one of the natural environments generally perceived as being most desirable to humans.[14] They are, by definition, pieces of land surrounded by water and thus geographically separate from the mainland. This separation, or 'otherness', often leads to islands and their inhabitants being imbued with all kinds of metaphors and meanings, including notions of isolation, insularity, mystery, connectedness, colonisation, romance, utopia, vulnerability, robustness and sanctuary. Before the advent of steam navigation, as Francoise Peron notes, journeys to islands were usually long and arduous, so there was often limited information about those places, fuelling the idea of them as 'other' or 'strange'.[15] Journeying to islands requires making a special effort; it involves the crossing of a threshold – the threshold of the sea. But perhaps journeying to an unusual island like Sark may also be compared with other journeys that involve crossing thresholds, whereby one's ordinary, known world is left behind and an unfamiliar, unknown

14 Yi-Fu Tuan, *Topophilia: A Study of Environmental Perception, Attitudes and Values*, 2nd ed., (New York: Columbia University Press, 1990), p. 247.

15 Francoise Peron, 'The Contemporary Lure of the Island', *Journal of Economic and Social Geography* 95.3 (2004): pp. 326–39, p. 328.

place is entered which has 'strange' ways.

This interest in islands has resulted in them being considered worthy of study in their own right. The relatively new academic field of Island Studies focuses particularly on four main areas: island governance, the future of islands, island tourism and the nature of islandness. Apart from looking at definitions and typologies of islands, topics explored under the nature of islandness include whether island communities may be considered special and distinctive, the characteristics of an insular cultural identity and the contrasting features of islands, which can be at the same time both culturally conservative and original.

Islands have also long fascinated writers. Gillian Beer, quoted in an article by Julia Bell, comments that, particularly in English culture, the island is the 'perfect form... just as the city was to the Greeks. Defensive, secure...a safe place'.[16] Unsurprisingly islands feature in some of that culture's best known literary classics, including William Shakespeare's, *The Tempest*, a remote island where sorcery takes place; William Goldings's *Lord of the Flies*, describing the disaster which befalls British boys stuck on an uninhabited island; and Enid Blyton's, *The Island of Adventure* set on the mysterious Isle of Gloom. Thus also illustrating that whilst an island can offer an individual a safe haven it can be a place of major personal upheaval and transformations

16 Julia Bell, 'Why Writers Treasure Islands: Isolated, Remote, Defended – They're Great Places for Story-telling', *Independent*, http://www.independent.co.uk/arts-entertainment/books/features/why-writers-treasure-islands-isolated-remote-defended-theyre-great-places-for-story-telling-10391558.html [Accessed 13 Oct 2015].

can occur.[17] This seemed to have been the case for Mervyn Peake who first visited Sark in 1933, when Eric Drake his former art teacher invited him to join an artists' colony there, and then came back after the Second World War to live with his family. He is remembered for getting his ears pierced on Guernsey, which was unusual at that time for men, and painting whilst naked on the headlands of Sark. In summer he picnicked and swam with his family at Grande Greve beach and used his drawings of them there as models for his *Treasure Island* illustrations. Clearly inspired by Sark, he wrote his gothic novel *Gormenghast* during his time there and it was also the setting for another novel, *Mr. Pye*. Sark's night sky could hardly escape Peake's notice, evidenced by lines such as, 'A full moon was flooding the island as though with phosphorus'.[18] Also, he clearly felt this was a place where he connected to all of nature as he illustrates in this poem, 'Life beat another rhythm':

> Life beat another rhythm on that island
> As old as her own birth
> We were the island people, and the earth
> Sea, sky, and love, were Sark, and Sark, the earth
> While round us moved the swarming of the sea.[19]

Thus, like many other writers before him Peake found an

17 Stephanos Stephanides and Susan Bassnett, 'Islands, Literature and Cultural Translateability', *Transtextes transcultures*, Hors série (2008), pp. 5–21: http://transtexts.revues.org/212 [Accessed 7 Jan 2016].

18 Mervyn Peake, *Mr. Pye*, (London: Vintage Books, 1999), p. 241.

19 Mervyn Peake, *Collected Poems* (Manchester: Carcanet Press, 2008), p. 70.

island could be a rich source of creativity and in Sark's case, lend itself well to his particular idiosyncrasies and brand of fantasy.

For small islands in particular with visible boundaries, living in such close proximity to the sometimes impassable sea can expand the sense of place and enhance the sense of connectedness to the natural world. It is difficult to ignore a storm on Sark, particularly when it results in the ferry to Guernsey being cancelled making travel to the mainland impossible. Equally, in springtime the smell of wild garlic, violets and bluebells from Dixcart Valley pervades the whole island. From the islanders point of view, not only is there often a strong sense of a shared distinct island identity and pride in being 'different' from those living on the mainland but, as Stephen Royle points out, islanders may also, 'develop a relationship with their island that becomes part of their identity'.[20] Henry Johnson suggests Sark can be viewed as a type of 'microstate' or 'nation' in its own right as, 'its degree of political autonomy and traditional practice offers an example of island identity within a sphere of concentric and interconnected political affiliations and power relationships'.[21] The British TV viewing public got a snapshot of one aspect of this microstate in the twelve episodes of the BBC Two documentary series, 'An Island Parish', which was broadcast between 2012–2014 – the series

20 Peron, 'Lure of the Island', p. 331; Stephen A. Royle, *Islands: Nature and Culture* (London: Reaktion Books, 2014), p. 55.

21 Henry Johnson, 'Sark and Brecqhou Space, Politics and Power', *Shima: The International Journal of Research into Island Cultures* 8.1 (2014): pp. 9–33, p. 10.

followed Sark's Anglican priest and Methodist minister and also featured a host of local people.[22] In 2014 when I began meeting people who were interested in taking part in my research, I could not quite understand why some of them seemed so familiar until I realised that I had already 'met' them on TV!

Although Sark may have its own unique way of life, it also shares certain characteristics with other small islands, including concerns regarding depopulation and a vulnerability to ecological, linguistic and cultural change from outside influences. It was settled during the Neolithic and Bronze Ages, and later by the Romans and Normans.[23] During the Second World War, Sark was occupied by German forces, and in 1990 it was subject to an attempt at invasion by a French unemployed nuclear physicist who believed himself to be the rightful Seigneur. Further, the Norman dialect *Sercquiais* (also known as *Sarkese*), a version of the dialect used by island's original colonists, has now almost completely died out.

In increasingly challenging economic times, islanders everywhere have had to adapt in order to survive. John Connell who has researched the political, economic and social development in small island states goes so far as to say, that such adaptations may involve the place they live in having to be, 'aesthesicised, sanitised and anaesthetised'

22 BBC Two, 'An Island Parish', http://www.bbc.co.uk/programmes/boo6t6m6 [Accessed 12 Mar 2016].

23 Victor Coysh, *Sark: The Last Stronghold of Feudalism* (Guernsey: Toucan Press, 1982), pp. 9–17.

and islanders then having to maintain a level of external compliance with the metaphors and meanings attached to their island in order to satisfy the demands of tourists.[24] Sark and its residents do not appear to escape some of these references as is illustrated by the island's description on a number of tourist websites as a, 'romantic hideaway', a 'mysterious island' and a 'paradise'. For those living and working on Sark however their everyday lives may reflect a different reality, there is a strong work ethic and many people are kept busy with multiple jobs. Thus, there is a subtle pressure on the island to be ageless and not subject to change. However such mythologising of Sark can be problematic as it can lead to stereotyping and the psychological projection of certain qualities and values onto not only the island itself but its residents too. An example of this was an exchange between a tourist and a local that I overheard on one of my visits: Tourist, 'So no point trying my mobile then?' Local, 'It'll work fine'. Tourist (disappointedly), 'Oh, I thought you didn't use them here…' It may be the case that Sark, with its potent combination of being not only a small island but also one with unusual customs and practices, leads to some people projecting onto the island the desire that the island meet their own needs about how the rest of the world should be.

Living on a small island also inevitably involves living in close proximity to other people, accommodating everyone's idiosyncrasies, tolerating differences in opinion and

24 J. Connell, 'Island Dreaming: The Contemplation of Polynesian Paradise', *Journal of Historical Geography* 29.4 (2003): pp. 554–82, p. 568.

accepting occasional disagreements and the development of factions. Many island communities can, as Peron points out, also be very territorial and quick to deal with perceived challenges from the outside.[25] Examples of this territoriality on Sark include the rejection in the 2008 election of the majority of candidates put forward by two billionaire brothers, Sir David and Sir Frederick Barclay who live on their neighbouring private island of Brecqhou in a mock-Gothic castle guarded by surveillance cameras and own the Telegraph Media Group – 90% of Sark's voters participated in the election. Territoriality is also demonstrated by the islanders' determination to maintain a 'no fly zone' over the island. Sark's relationship with the nearby even smaller island of Brecqhou, owned by the Barclay Brothers, is complex and the situation can be viewed as a 'contested geography', in the sense that, as Johnson comments, between the two islands, 'the notion of island space is territorialized and challenged'.[26] This is despite Sark's historical jurisdiction over Brecqhou already being confirmed and Brecqhou becoming a 'tenement' (plot of land) of Sark in 1929. Reflecting on this relationship I was reminded of a binary star system where two stars are in such close proximity to each other that every so often the stellar atmosphere that they share is altered. Perhaps it is inevitable that events on one island will influence the other, at its nearest point Brecqhou is after all only 70m from Sark, separated by the narrow Gouliot channel.

25 Peron, 'Lure of the Island', p. 330.
26 Johnson, 'Sark and Brecquou', p. 12.

THE JOURNEY TO BECOMING A DARK SKY COMMUNITY

Sark's night skies have always been recognised by residents as being very dark, and they play a major role in contributing to the island's uniqueness. This extreme, and for many unfamiliar, darkness is often one of the first things visitors comment on. In 2009, Jo Birch, of Sark Tourism and *La Société Sercquaise*, (a local society, 'founded to study, preserve and enhance Sark's natural environment and heritage' which also supports astronomy) heard that the sparsely populated Galloway Forest Park in southwest Scotland had become Europe's first Dark Sky Park and that this had generated a huge amount of publicity for Galloway, both in the UK and throughout the world. Birch then decided to contact Steve Owens, the astronomer and dark skies consultant who had co-led the Galloway IDA application to ascertain if Sark could embark on a similar process. This resulted in Birch, along with a small group of others on Sark, setting into motion the application process for Sark to become an IDA-recognised Dark Sky Community.

Owens acknowledges that embarking on such a time consuming, complex application process requires the passion and commitment of a small group of advocates active in their local community who know the area well and that such initiatives work best when they are 'bottom-up' rather than 'top-down'.[27] It was also recognised that not only the

27 Steve Owens, 'Astronomical Tourism in Dark Sky Places', www.wcmt.org.uk/sites/default/files/migrated-reports/952_1.pdf [Accessed 12 Dec 2015].

commitment of those individuals but the commitment of the whole island would be required for a successful outcome. The Tourism and Agriculture Committees of the Chief Pleas, which also oversees environmental matters, were supportive of the application, recognising the conservation and economic values in having Sark awarded Dark Sky Community Status, and they granted £2000 towards the lighting management plan and associated costs. Sark Chamber of Commerce were also in favour. An appeal was generated by *La Société Sercquaise*, to help fund the application and this resulted in a further £2,725 being raised by the local community. Many people also contributed in kind, providing free accommodation, bicycles and trailer hire, etc. further demonstrating strong community will for the application to succeed.

Work then began with Owens to construct a plan to further the process. The plan involved five stages:

1. Carry out an initial dark sky survey of the skies above Sark, to quantify how dark they are.

2. Carry out a full lighting audit of all external lights on Sark, to estimate the percentage of lights that were compliant with good lighting practice.

3. Commission a Lighting Management Plan from a member of the Institute of Lighting Engineers, such that the plan can be formally adopted by the Agriculture and Environment Committee of the Chief Pleas and used to ensure good lighting practice across the island.

4. To identify those lighting fixtures that fail to comply with good lighting practice and replace or modify them to ensure at least a 75% compliance rate.

5. Produce and submit to the IDA an application pack to become a Dark Sky Community.[28]

Although Philip's Dark Skies Map, which indicates the visibility of stars from any location in Britain and Ireland, is useful in showing how small an impact light pollution from surrounding areas has on Sark, the IDA required this to be quantified.[29] When Owen used a Sky Quality Meter to measure sky brightness at a number of sites on the island, Sark's night sky was found to have a class of 3 on the 9-level Bortle Dark-Sky Scale where the lower the number the darker the sky, 1 is an utterly unpolluted sky and 9 is an inner-city sky. At Bortle class 3 sites it is usual to be able to see around 4,500 stars with the naked eye. This proved what people on Sark already knew – their sky was very dark!

The assessment process took a year of hard work and involved seeking the approval of the whole community. There was overwhelming assent. Jim Patterson of the Institute of Lighting Engineers developed a comprehensive lighting management plan, and over a period of a week Owens visited every outside light on the island confirming the sky's extreme darkness. In an interview with Chan-

28 Owens, 'Sark Dark Sky Community. A Dark Sky Island'.

29 Philip's Astronomy, *Philip's Dark Skies Map Britain and Ireland: Darkest Observing Sites in the British Isles* (London: Philip's Astronomy, 2004).

nel News he said, 'to all intents and purposes you've got skies darker than ninety nine percent of people in the UK will ever see'.[30] His audit took longer than he had expected as the enthusiasm of many of the locals meant he was often stopping to chat. Only two individuals declined to have their lighting measured and in total only around ten household lights needed to be changed. The lighting management plan assisted local residents and businesses in making changes to their lighting to ensure that as little light as possible escaped upwards or spilled out. Several of the hotels and Sark School also refitted their lighting; the Aval du Creux Hotel at the top of Harbour Hill had all 65 of its external lights refitted as downlighters. The application to the IDA contained a number of letters of support from Sark residents, including Michael Beaumont, the Seigneur at that time (now deceased), Sandra Williams, Chair of Sark Tourism Committee of the Chief Pleas, Sarah Cottle, Head teacher, Sark School and Jeremy La Trobe-Bateman, then Chair of the Sark Astronomy Group (a forerunner to SAstroS). Jeremy's own family have a long association with Sark. In 1914 two of his ancestors, brothers Geoffrey and Leslie La Trobe Foster wrote *La Trobe Guide to the Coasts, Caves and Bays of Sark* which Jeremy and Rob Pilsworth (a Sark enthusiast) revised and reissued in 2014 to mark the *Guide*'s centenary.[31] The application was also supported by

30 Steve Owens, 'Dark Skies on Sark', https://www.youtube.com/watch?v=_HUmuwCrbPo [Accessed 12 Dec 2015].

31 Geoffrey and Leslie La Trobe, *La Trobe Guide to the Coasts, Caves and Bays of Sark*, 7th rev. ed., eds. Jeremy Latrobe-Bateman and Rob Pilsworth (Sark: Lazarus Publications NFP, 2014).

business people such as Paul Armorgie, Director of Stocks Hotel, and Sally and Peter Hutchins, owners of Vieux Clos Guest House. Sark Electricity Company Ltd. (a privately owned company), interestingly, were also in favour.

The reasons behind the government and community's drive to apply for Dark Sky status seem to have been primarily economic. During the summer months, Sark's population expands from around 550 to over 40,000. The 'shoulder season' (between October to March) is usually much less busy and many hospitality businesses on the island close then. Sark Tourism, operated by the Tourism Committee of the Chief Pleas, would like tourism numbers to increase during these months. They believe that if visitors could be encouraged to see that Sark held other attractions during the autumn and winter months, then more income could result. This was a view also shared by Owens: 'This an ideal opportunity to bring stargazers to the island throughout the year, and I think that Sark is about to see a boom in astro-tourism, especially in the winter months'.[32] Also, this opens the potential to educate locals and visitors about astronomy, the benefits of preserving dark skies and of keeping outdoor lighting to a minimum. To this aim Sark Tourism produced and sent out two leaflets to all residents on Sark, 'Keeping Sark Dark Sky Friendly' and, 'Property Self Audit Guidelines'.

Some dark sky places, such as Galloway Forest Dark Sky Park and Kerry International Dark Sky Reserve, faced problematic issues such as where to set the beginning and

32 Owens, 'Dark Skies on Sark'.

end of the 'Core Zone' and the 'Buffer Zone'. As Sark is an island, and the IDA application referred to the whole island, these potentially problematic issues did not apply. On Sark the 'Core Zone' is the whole island and the 'Buffer Zone' is simply the English Channel! This also made the monitoring of the Lighting Management Plan much more straightforward.

In 2011 the application was approved and Sark became a Dark Sky Community and the world's first dark sky island. Conseiller Paul Williams, chair of the Agriculture Committee said, 'Sark becoming the world's first dark sky island is a tremendous feather in our environmental cap, which can only enhance our appeal'.[33]

33 Sark Chief Pleas, 'Sark Hailed as the World's First Day Sky Island', http://www.socsercq.sark.gg/News%20and%20Projects/darkskiespressrelease.html [Accessed 7 Dec 2015].

International dark sky places

International Dark Sky Communities are one of the five designations awarded by the IDA, the four other designations being:

- International Dark Sky Parks

Parks are publicly or privately owned spaces protected for natural conservation that implement good outdoor lighting and provide dark sky programs for visitors.

- International Dark Sky Reserves

Reserves consist of a dark 'core' zone surrounded by a populated periphery where policy controls are enacted to protect the darkness of the core.

- International Dark Sky Sanctuaries

Sanctuaries are the most remote (and often darkest) places in the world whose conservation state is most fragile.

- Dark Sky Developments of Distinction

Developments of Distinction recognize subdivisions, master planned communities, and unincorporated neighborhoods and townships whose planning actively promotes a more natural night sky but does not qualify them for the International Dark Sky Community designation.

To date there are 69 Dark Sky Places in total, with the majority located in the US. Coll, in the Inner Hebrides, Scotland, became the world's second Dark Sky island in 2013.

DARKSKY.ORG/IDSP

THE BIRTH OF THE SARK ASTRONOMY SOCIETY (SAstroS)

At the time of the IDA application in 2010, Sark was home to a small number of amateur astronomers, but there were no regular organised astronomical events on the island. Felicity Belfield, a long-term Sark resident, noted however, 'Considering the population of Sark; approximately 600, there are a lot of telescope owners and much interest in astronomy'.[34] Although I heard anecdotally from locals that sometimes the telescopes are used for spotting boats rather than stars! The IDA application required community promotion of dark sky education, and so the *Sark Astronomy Society* came into being soon after the award of Dark Sky status following a suggestion by the IDA. SAstroS was formed by a group of local stargazers who wanted to share Sark's magical night sky with others. In an interview with *Outdoor Nation* in March 2013, Annie Daschinger, Chair of SAstroS (also affectionately known as ' the Starfleet Commander'), was keen to stress that, 'None of us in SAstroS is an "astronomer", although we are gradually learning and deeply appreciative of our unique status, (and) we see our role as primarily to provide information and access for the purpose of bringing astro-tourism to Sark'.[35]

34 Felicity Belfield, in Owens, 'Sark Dark Sky Community: A Dark Sky Island'.

35 *Outdoor Nation*, 'Q&A with Annie Daschinger and Jo Birch: The World's First 'Dark Sky' Island', http://blog.outdoornation.org/qa-with-annie-dachinger-and-jo-birch-the-worlds-first-dark-sky-island/ [Accessed 7 Dec 2013].

Since its small beginnings SAstroS now has more than thirty members, a Facebook page with 363 members (October 2016) and has run a number of public events biannually in the autumn and spring. Their most recent endeavour was the Autumn *Starfest* which took place in October 2015, but this time it ran alongside a new event, *Sark Trek*, organised in conjunction with Stocks Hotel, Sark and Cosmos Planetarium. *Sark Trek* is a new type of stargazing weekend break, designed to attract people with little or no knowledge of astronomy. It is a measure of the growing awareness of Sark's uniqueness that such events attract well-known astronomers such as Dr. Marek Kukula, Public Astronomer at the Greenwich Royal Observatory, and Professor Chris Lintott, University of Oxford and presenter of BBC's *The Sky at Night*.

SAstroS' latest endeavour, after more than a year of exploring three different possible sites, fundraising and planning, has been to build a small observatory situated near The Mill in a high position in a field in the middle of the island, one of Sark's darkest spots. This houses their second-hand ten-inch telescope acquired in 2014, funds for which were raised in just a few days, and some smaller telescopes, star charts and books. The observatory is a custom-built, wooden one-storey building, with one room for the telescopes (with sliding roof) and an adjoining 'warm' room, which, once additional funds are raised, will house a screen connected to the telescope to ensure stargazers will not get cold. It was pre-fabricated in Norfolk by a company called Home Observatory UK. The £10,000 funding required was raised by SAstroS but, as Daschinger says, 'None of this could have happened without the support of Sark residents and the generosity of members of SAstroS'.[36] When I visited the observatory site in September 2015 the foundations were in place. A local resident told me she was storing the sign for the observatory in her house, excitedly awaiting its imminent arrival. As Sark has no airport it would arrive, via Poole and Sark Shipping, on the 29th September and be built by the 4th October. Like everything of any size arriving on the island, it would be transported up Harbour Hill by tractor. And indeed, on schedule, the observatory was opened by Kukula on the weekend of SAstroS' Autumn *Starfest* and a public open afternoon followed soon after. Tours of the observatory can be arranged through Sark Tourism.

SHINING THE LIGHT: FOCUSING THE RESEARCH

'Holes in the sky, says the child scanning the stars'; this line of Louis MacNeice's poem, 'Holes in the Sky', has always intrigued me.[37] It seems to suggest there are different ways to interpret the wonder of what we see in the night sky. When I first began thinking about my research topic, I found myself remembering this poem. I also recently discovered that

36 BBC News, 'Sark, the first Dark Sky Island, Gets Observatory', http://www.bbc.co.uk/news/world-europe-guernsey-32596199 [Accessed 4 Nov 2015].

37 Louis MacNeice, *Holes in the Sky: Poems, 1944–1947* (London: Faber and Faber, 1948).

in Native American mythology there is a story of how the night sky came about, how the hummingbird made 'holes in the sky' in an effort to get the sun back and reconcile the animals. Whereas I can now, with my basic level of astronomical knowledge, recognise and name some stars, as a child I was content to simply look up with curiosity at those bright 'holes'.

To begin establishing a broader foundation for my research I initially searched within the fields of ecopsychology, health and environmental psychology, and cultural astronomy as these were the fields which seemed to have most to say regarding the human relationship with nature and the environment. I also dipped into works which explored the long-standing fascination humankind has with the night sky, including cultural histories of darkness and personal testimony. Later, I expanded my search to include various forms of tourism such as astronomical tourism, heritage tourism and nostalgia tourism, as I believed doing so could help me understand more about what might motivate people to visit Sark in the first place. My investigation focused on exploring the following themes: the human desire to see the night sky, the commercialisation of this desire through astronomical tourism, fear of the dark, the 'nature' of nature, and nature and wellbeing.

Although a number of ecopsychologists and environmentalists have suggested that encounters with nature and encouraging a human/nature connection may be transformative and beneficial to the wellbeing of individuals and communities, their focus tends to be on 'green'/earth-bound nature rather than encounters with the sky.[38] In addition, there have been attempts to not only describe but also measure the transpersonal dimensions to nature experiences such as feelings of identifying with the environment, of oneness and unity, changes in perception of time while in a natural location and any changes in personality reported as a result of peak and transformative experiences.[39] Some people associated with the Dark Sky movement have however suggested that observing the night sky with the naked-eye (rather than a telescope) may also result in similar transformative and beneficial effects and this is the area I particularly wanted to focus on to see if others had also experienced what I had when out at night on Sark.[40]

There is no one simple definition of what constitutes 'nature' and a search through the literature uncovers a

38 *Ecopsychology: Restoring the Earth. Healing the Mind*, eds. Theodore Roszak, Mary E. Gomes and Allen D. Kanner (San Francisco: Sierra Club Books, 1995); Terry Hartig, Marlis Mang and Gary W. Evans, 'Restorative Effects of Natural Environment Experiences'. *Environment and Behavior* 23.1 (1991): pp. 3–26; Rachel Kaplan and Stephen Kaplan, *The Experience of Nature: A Psychological Perspective* (Cambridge: Cambridge University Press, 1989); Roger S. Ulrich, 'Aesthetic and Affective Response to Natural Environment', in *Human Behavior and Environment: Advances in Theory and Research. Vol. 6: Behavior and the Natural Environment*, eds. I. Altman and J.F. Wohlwill (New York: Plenum Press, 1983).

39 Freya Mathews, *The Ecological Self* (London: Routledge, 1991), p. 162; K. Williams and D. Harvey, 'Transcendent Experience in Forest Environments', *Journal of Environmental Psychology* 21.3 (2001): pp. 249–60.

40 Bogard, *End of Night*; Julie James, 'Fifth International Dark Sky Reserve Designated in Wales', http://www.darksky.org/night-sky-conservation/283 [Accessed 29 Oct 2013].

wealth of material on the subject.[41] Within the fields of ecopsychology and health and environmental psychology, references to nature do not usually specifically include celestial bodies and sky features. Within recent times, the sky has not been considered to be part of nature. Writing about the sky in the early twentieth century, Victorian art critic John Ruskin described this lack of attention:

> Sky is the part of creation in which Nature has done more for the sake of pleasing man, more for the sole and evident purpose of talking to him and teaching him, than in any other of her works, and it is just the part in which we least attend to her.[42]

This 'missing sky factor' is illustrated by descriptions of nature as comprising, 'plants, objects (such as rocks), events (such as storms), and of course animals', the classification of nature as, 'soils as well as plants and animals and their supporting habitats' and the categorising of natural environments into subcategories of 'totally natural', 'civilised

41 N. Evernden, *The Social Creation of Nature* (Baltimore: Johns Hopkins University Press, 1992); P. Macnaghten, and J. Urry, *Contested Natures* (London: SAGE Publications, 1998).

42 John Ruskin, *Of the Open Sky*, Modern Painters I, Part II, Section III, http://www.lancaster.ac.uk/depts/ruskinlib/Modern%20Painters [Accessed 12 Mar 2016].

natural', 'quasi-natural', 'semi-natural' and 'non-natural'.[43] However the IDA considers the sky to be, 'one half of the entire planet's natural environment'.[44]

Until the early nineteenth century, it was common practice in the UK to depend on the moon and stars for light, particularly in rural areas. However the explosion since then in the use of artificial light in the environment has meant these celestial bodies are no longer our primary sources of light at night, and natural darkness has become an alien medium. The amount of darkness permissible in different settings has become increasingly monitored and managed. Although there are benefits to having adequate light at night, including increasing the amount of time available for work and leisure, sometimes the type or amount of light is not appropriate or inefficient and can be considered to have a polluting effect. Light pollution comes in various forms but is generally accepted to come under three main headings: glare (from unshielded or badly directed lights that shine horizontally leading to the subject being obscured rather than illuminated); sky glow (the glow or halo effect over populated areas at night caused by reflected and unshielded light reaching the sky); and

43 Peter Kahn and Stephen Kellert, *Children and Nature: Psychological, Sociocultural, and Evolutionary Investigations* (Cambridge, MA: MIT Press, 2002), p. xiii; Linda Seymour, *Nature and Psychological Wellbeing. English Nature Research Report Number 533* (Peterborough: English Nature, 2003), p. 7; Claudia Mausner, 'A Kaleidoscope Model: Defining Natural Environments', *Journal of Environmental Psychology* 16.4 (1996): pp. 335–48.

44 International Dark-Sky Association, 'What We Do', http://www.darksky.org/about-ida [Accessed 7 Jan 2014].

light trespass or light nuisance (when light is too intense or far reaching and may encroach on another's property). Over-illumination is a form of light trespass that occurs when light is kept on when it is not needed, e.g., empty office or university buildings lit up all night.[45] The sharpest increase in light pollution across the UK was between the mid to late 1980s and it is estimated that it is still growing at a rate of 3% annually; from 1993 to 2000 light pollution in England alone increased by an astonishing 26% with a higher rate of increase recorded in many areas of the US.[46] Since 2007 Globe at Night has been running an international citizen-science campaign to measure night sky brightness whereby members of the public submit their observations by computer or smart phone.[47] So far people in 115 countries have contributed 100,000 readings mapping ever-encroaching light pollution. Naked-eye astronomers all over the world have become increasingly aware of the negative effects of nocturnal sky glow particularly in urban areas, and there is virtually no professional astronomy now taking place in the UK at all (although the weather is also partially to blame).

The Campaign for Dark Skies (CFDS) is a branch of the British Astronomical Association (BAA) and is the main UK

45 Globe at Night, 'What is Light Pollution?', http://www.globeatnight.org/light-pollution.php [Accessed 2 Feb 2016].

46 CPRE, 'Campaigning for a Starry Starry Night', http://www.cpre.org.uk/magazine/opinion/item/3737-campaigning-for-a-starry-starry-night [Accessed 2 Feb 2016].

47 Globe at Night, 'About Globe at Night', http://www.globeatnight.org/about.php [accessed Feb 2 2016].

body lobbying against light pollution. The CFDS originally arose from amateur astronomers becoming concerned that the increase in artificial lighting was impeding their views of the night sky but came to recognise light pollution as an issue affecting many people. For professional astronomers light pollution may be less of an area of concern as they are involved with data interpretation and generally not viewing from public spaces or their own houses.[48] Between 2009/10 The Campaign to Protect Rural England (CPRE) and the CFDS conducted a Lighting Nuisance survey. The survey found that the problems light pollution caused included 83% of respondents reporting their view of the night sky was affected with 50% stating light shining in to a bedroom window hindered their sleep.[49] A variety of measures had been considered to deal with the problem including confronting the person responsible for the light pollution or, in a small number of cases, moving house.[50] 71% had however not complained and those who had complained to their local authority found them largely unsupportive. A number of cities in the UK however (including Edinburgh) have now begun to implement strategic lighting master plans whose purpose is to ensure all public lighting is co-ordinated and thereby prioritise 'the visual experience of the city for pe-

48 Oliver Dunnett, 'Contested Landscapes the Moral Geographies of Light Pollution in Britain', *Cultural Geographies* 22.4 (2015): pp. 619–36, p. 621.

49 CPRE, 'Lighting Nuisance Survey 2009/10: Report', http://www.cpre.org.uk/resources/countryside/dark-skies/item/1974-lighting-nuisance-survey-2009-10-report [Accessed 12 Dec 2015].

50 CPRE, 'Lighting Nuisance Survey'.

destrians'.[51] Generally however when local authorities are looking at reducing artificial light in the name of reducing carbon impact or improving sustainability the rationale is often financial rather than environmental.[52] In the 1960s and '70s, as a child growing up in a new town in the west of Scotland, my bedroom each night was full of trespassed light from the orange sodium vapour light that stood right outside my window. These lights also lined the nearby main road which connected my town to Glasgow, but no-one ever questioned why it was necessary to have so many lights and the cost implications of so much of that light being scattered upwards.

Since the 1980s, a growing number of other local, national and international campaigns and organisations, such as the United Nations Educational, Scientific and Cultural Organisation's (UNESCO) Starlight Reserves programme and The World at Night (TWAN), have emerged in an attempt to encourage the reduction of light pollution and improve views of the stars.[53] More and more areas are choosing to become involved in the IDA's Dark Sky Places programme and the Association is particularly keen to encourage grassroots initiatives in areas which are not necessarily seeking Dark Sky status but are nevertheless

51 Royal Commission on Environmental Pollution, *Artificial Light in the Environment*, (London: The Stationery Office, 2009), p. 12.

52 Robert Shaw, 'Streetlighting in England and Wales: New Technologies and Uncertainty in the Assemblage of Streetlighting Infrastructure', *Environment and Planning* A 46.9 (2014): pp. 2228–42.

53 IDA, 'What We Do', Starlight Initiative, Welcome to the Starlight Universe', http://www.starlight2007.net/ [Accessed 9 Feb 2014].

aiming for balancing the need for appropriate lighting in their communities with preserving darkness. Although the IDA emphasises the importance of preserving culture and heritage stating, 'a lost view of the stars extinguishes a connection with the natural world', its literature, and that of the other Dark Sky initiatives, suggests its focus of interest is primarily on encouraging the use of environmentally responsible outdoor lighting.[54] Little attention appears to be paid as to how exactly culture and heritage can be preserved and the potential positive impacts on human wellbeing of doing so. Arguing for dark skies Bogard offers a reminder of the cultural losses that are at risk of being lost for ever, 'night's ancient gifts of quiet peace and time to be with those we love'.[55] The IDA also recently commissioned an article from Bogard on the night sky's inspiration and influence in literature which, to date, has not yet been published.[56]

Although there is a large amount of literature that talks about the value of dark skies, I discovered that there was a lack of any research regarding possible benefits to wellbeing. One exception to this deficit is the increasing amount of attention being paid to the negative impact of light at night on human physical health (including immune system functioning, disease, cancers and sleep disorders). In my work as a counsellor for students at a large Scottish uni-

54 IDA, 'What We Do'.

55 Paul Bogard, *Let There be Night: Testimony on Behalf of the Dark* (Nevada: University of Nevada Press, 2008), p. 5.

56 HarperCollins, 'Paul Bogard', http://www.harpercollins.co.uk/authors/11241/paul-bogard [Accessed 25 Jan 2014].

versity, I have noticed a big jump over the last few years in the number of students presenting with problems related to sleep. It is no coincidence that many of them are working and sleeping in the same room using laptops, smart phones and e-readers, all of which emit light on the blue wavelength. Increasingly it is being recognised that this 'blue light' suppresses the hormone melatonin which is secreted at night and assists the body's biological clock, and consequently interferes with sleep patterns.[57] Unfortunately, it is also now understood that the light-emitting diode (LED) outdoor lights, gradually replacing the old sodium lights as they are more energy efficient and provide the same visibility but with less brightness, have significant levels of 'blue light'.[58] Also, the LED 'retrofit revolution' taking place in many towns and cities appears to be leading to many areas spending the money saved on decreased electricity use by buying more lighting.[59] And yet in Europe before the advent of domestic electric lighting and the eight hour working day it was the norm for people to have a segmented sleeping pattern, going to bed at sundown for a 'first sleep', waking for an hour or so when they might engage in various other activities, and then having a 'second sleep' till

57 Harvard Health Publications, 'Blue light has a dark side'http://www.health.harvard.edu/staying-healthy/blue-light-has-a-dark-side [Accessed 12 Dec 2015].

58 Harvard Health, 'Blue Light'.

59 International Dark Sky Association, '5 Popular Myths About LED Streetlights', http://darksky.org/5-popular-myths-about-led-streetlights/ [Accessed 12 Dec 2015].

dawn.[60] It may be that our commonly experienced 'middle-of-the-night insomnia', is actually a leftover from this previous sleeping pattern.

But it is not only humans that need darkness, there is also a growing body of research looking at the effects of light pollution on the behavioural patterns (including breeding, feeding, attraction to light and migratory behaviour) and safety of nocturnal animals.[61] In fact there has been more research devoted to the ecological consequences on flora and fauna than to effects on human health. Bogard captures the devastation ecological light pollution wreaks when he describes it as, 'like the bulldozer of the night'.[62] Birds are a particularly affected as unnecessarily illuminated communication towers and high rise buildings may

60 Ekirch, *At Day's Close*, p. 300.

61 David Blask, George Brainard, Ronald Gibbons, Steven Lockley, Richard Stevens, and Mario Motta, 'Adverse Health Effects of Nighttime Lighting. Comments on American Medical Association Policy Statement', *American Journal of Preventive Medicine* 45.3 (2013): pp. 343–46; Michele Blackburn, Curtis Burney and Louis Fisher, *Management of Hatchling Misorientation on Urban Beaches of Broward County, Florida: Effects of Lighting Ordinances and Decreased Nest Relocation* (Broward County, FL: Environmental Protection Department, 2007); Franz Hölker, Timothy Moss, Barbara Griefahn, *et al.*, 'The Dark Side of Light: A Transdisciplinary Research Agenda for Light Pollution Policy', *Ecology and Society* 15.4 (2010): A13; Charlotte Bruce-White and Matt Shardlow, *A Review of the Impact of Artificial Light on Invertebrates* (Peterborough: Buglife – The Invertebrate Conservation Trust, 2011); Royal Commission on Environmental Pollution, *Artificial Light*; Angela Spivey, 'Light at Night and Breast Cancer Risk Worldwide', *Environmental Health Perspectives* 118.12 (2010): A525.

62 Paul Bogard, *Los Angeles Times*, 'Let There Be Dark', http://articles.latimes.com/2012/dec/21/opinion/la-oe-bogard-night-sky-20121221 [Accessed 12 Nov 2015].

affect their flight path. In 1981 more than ten thousand birds were injured or killed when they flew into floodlit smokestacks at the Hydrox Generating Plant near Kingston, Ontario.[63] Whilst the field of photobiology studies the influence of light on living organisms, in more recent years, a number of scientists and researchers from other fields have begun to recognise the lack of attention paid to the biological systems which require darkness to function. To give some idea of the huge numbers of animals involved, nocturnal animals comprise the vast majority of small rodents and carnivores, four-fifths of marsupials and a fifth of all primates.[64] In 2003, at a symposium entitled, 'Ecology of the Night: Darkness as a Biological Imperative', a new field of science, scotobiology, began to emerge. Its scope is wide and includes not only the biological effects of light pollution on animals, birds, insects, plants and humans but also the sociological, anthropological and cultural aspects of light pollution. Already those interested in this field have identified a large number of diverse areas that could benefit from further research. These range from exploring the effects of light pollution on the mental health of humans (and specifically humans from various racial and cultural backgrounds) to calculating the monetary value for society if workers' health and productivity could be increased, to recognising the high economic value attached

63 Sharon Guynup, 'Light Pollution Taking Toll on Wildlife, Eco-Groups Say', *National Geographic News*, http:// news.nationalgeographic.com/.../04/0417_030417_tvlightpollution.html [Accessed 12 Feb 2015].

64 Ron Chepesiuk, 'Missing the dark', *Environmental Health Perspectives* 117.1 (2009): A20–27, http://ehp.niehs.nih.gov/117-a20/ [Accessed 12 Feb 2015].

to many of the species affected by light pollution, who are either viewed as aesthetic objects (on game reserves) or for farming and hunting.

Whilst dark sky supporters such as Julie James, Chair of Brecon Beacons National Park Authority, claimed that there are, 'wellbeing benefits attached to this wonderful accolade, (*attaining International Dark Sky Reserve status*),' she did not offer any evidence to back up this claim.[65] Like most other people she seems to take it for granted that that this must be the case. In carrying out my research I hoped therefore to not only address the 'missing sky factor' in the existing research but also strengthen the Dark Skies movement's claims that dark night skies can have a positive impact on wellbeing.

65 James, 'Dark Sky Reserve'.

CHAPTER 2:

SCANNING THE SKIES: IDENTIFYING THE THEMES

THE HUMAN DESIRE TO SEE THE NIGHT SKY

EVERY CULTURE APPEARS to have paid attention to the sky, and as Nicholas Campion remarks, 'there is no human society that does not, somehow, in some way, relate its fears, concerns, hopes, and wishes to the sky'.[1] Over the centuries artists, writers and musicians all over the world have found inspiration in starry night skies creating paintings such as, 'The Starry Night' by Vincent Van Gogh, 'The Planets' an orchestral suite by Gustav Holst, and more recently songs such as, 'Starlight' by Angharad, the official Dark Sky Reserve song for the Brecon Beacons National Park and 'Dark Skies', by musician Emma Pollock. Pollock's inspiration for her song came from childhood memories of

1 Campion, *Astrology and Cosmology*, p. 1.

being in Galloway Forest Dark Sky Park, 'staring up at the stars in the quiet of the forest'.[2] In 2014 a twenty-four hour experimental music and art event, 'Sanctuary' took place in the Park. Created by local Galloway artists Robbie Coleman and Jo Hodges in collaboration with many other artists, the event included light installations, sound works and radio transmissions all celebrating darkness. Sark's night skies have also stimulated creativity. During his visit to Guernsey in 1832, Joseph Mallord William Turner did a number of pencil sketches of Sark's high jagged cliffs including, *La Coupée, Sark Island*.[3] In 2011 twenty artists from the Artists for Nature Foundation visited Sark on two occasions as part of the project 'Art for the Love of Sark'. The project came about due to the efforts of Sarkee and wildlife artist Rosanne Guille who had concerns for the future sustainability of the island and approached the Foundation. The artists recorded all aspects of island life in painting and sculpture. One artist in particular, Vadim Gorbatov from Russia, completed two paintings of the Sark night sky, 'Milky Way' and 'Evening, Bats'. The project resulted in the publication of a book, *'Art for the Love of Sark' A Contemporary Portrait of a Changing Island* and exhibitions.[4] Also, in 2015, the singer Enya released an album, *Dark Sky Island*

2 *M magazine*, 'Interview: Emma Pollock', http://www.m-magazine.co.uk/features/interviews/interview-emma-pollock/ [Accessed 12 Feb 2016].

3 Tate, 'Joseph Mallord William Turner, "La Coupée, Sark Island", ?1832', http://www.tate.org.uk/art/artworks/turner-la-coupee-sark-island-d23637 [Accessed 12 Mar 2016].

4 Chris Andrews, Renate Zoller and Amy McKee eds., *Art for the Love of Sark*, (Oxford: Gateway Publishing Ltd., 2012).

which came out of her own experiences of making both physical and emotional journeys. One of the songs is also called 'Dark Sky Island' and was inspired by and named after Sark (although to date she has not apparently actually visited Sark!)

This enduring desire to observe the night sky is fostered by organisations such as For Spacious Skies, whose founder Jack Borden observes, 'when you realize that everyone is in the sky instead of under it, as many people perceive themselves, you get a stronger sense of connectedness'.[5] Appreciating the night sky may also be a means of connecting profoundly with the past, as for example when Joe Slovick, speaking of Native Americans, describes experiencing, 'the very same view of the sky that was seen by the Chacoans a thousand years ago'.[6] Although we generally see the same constellations our ancestors did due to the stars appearing to be fixed in relation to each other, it is getting harder to spot them in many areas. To better comprehend our place in the universe and get a sense of our planet's physical insignificance we need to be able to see the stars. Yet with increasing light pollution worldwide, this experience is becoming much less common for many people, particularly

5 Jack Borden, 'For a New View of the World: Sky Walk', http://www.prevention.com/fitness/fitness-tips/reduce-stress-sky-walking [Accessed 23 Dec 2013].

6 Joe Slovick, 'Towards an Appreciation of the Dark Night Sky', http://www.georgewright.org/184sovick.pdf [Accessed 2 March 2014].

in North America and Western Europe.[7] In 1994 when an earthquake resulted in a massive power outage in Los Angeles, worried residents rang the authorities about a huge, silvery cloud. This was the Milky Way, which many had never seen before due to sky glow.[8] Unfortunately the view Van Gogh had of the Milky Way from the window of his sanatorium in Saint-Rémy-de-Provence, France can also no longer be seen.

Appreciation of the night sky may also lead to a more intense attachment to it. The concept of *noctcaelador* – strong interest in, and attachment to, the night sky – was first investigated by William Kelly who developed the *Noctcaelador* Index in an attempt to measure individual differences in people's psychological attachment to the night sky. Higher scores indicate a stronger attachment and interest.[9] Research was carried out with 150 US college students who were asked how regularly they intentionally looked at the night sky and to what extent opportunities to see the night sky would influence decisions in choosing where they

7 Dark Skies Awareness, 'Light Pollution – What is it and Why is it Important to Know?' http://www.darkskiesawareness.org/faq-what-is-lp.php [Accessed 12 Feb 2015].

8 Andrew Fraknot, 'Light Pollution', http://www.pbs.org/seeinginthedark/astronomy-topics/light-pollution.html [Accessed 12 Feb 2016].

9 William E. Kelly, 'Development of an Instrument to Measure *Noctcaelador*: Psychological Attachment to the Night-Sky', *College Student Journal* 38.1 (2004): pp. 100–2; William E. Kelly, 'Night-sky Watching Attitudes among College Students: A Preliminary Investigation', *College Student Journal* 37.2 (2003): pp. 194–96, William E. Kelly, and Kathryn E. Kelly, 'Further Identification of *Noctcaelador*: An Underlying Factor Influencing Night-Sky Watching Behaviors'. *Psychology and Education: An Interdisciplinary Journal* 40.3–4 (2003): pp. 26–27.

might live, whether they miss sleep before an exam and any memories they had of watching the night sky in childhood. Kelly concluded that *noctcaelador* does influence night-sky watching behaviours and attitudes; those who felt an attachment to the night sky chose to do so over other behaviours.[10] In a study of sky knowledge around the world, Jarita Holbrook employed the *Noctcaelador* Index to examine whether increasing light pollution might be a factor in the apparent decline in knowledge about the night sky, and, in addition, if this appreciation and knowledge is declining because people generally now spend more time indoors at night.[11] Holbrook expected that those raised in less light-polluted environments would record higher scores on the Index, thereby suggesting a stronger psychological attachment to the night sky, however in a survey of a hundred people who had been brought up in a variety of physical environments, Holbrook's results were not statistically significant.[12] More research is needed to test the Index's validity, which could incorporate not only self-reporting but also observable behaviours to further describe how this attachment to the night sky may manifest.[13]

With regard to the many stories, myths and meaning pertaining to the night sky, Anthony Aveni comments that

10 Kelly, 'Development', p. 102.

11 Jarita Holbrook, *Sky Knowledge, Celestial Names, and Light Pollution* (Unpublished MS, University of Arizona, 2009).

12 Holbrook, *Sky Knowledge*, pp. 3–4.

13 William E. Kelly and Jason Batey, 'Criterion-Group Validity of the *Noctcaelador* Inventory: Differences Between Astronomical Society Members and Controls', *Individual Differences Research* 3.3 (2005): pp. 200–3, p. 202.

our ancestors used their imagination to, 'mould a wonderful poetic imagery about themselves and their relationship to the universe'.[14] John D. Barrow also observes many of the myths generated are 'often attempts to join the heavens and the Earth'.[15] Those myths parallel some of the material from ecopsychology in that the sky is considered to be part of nature and connection with it seen as valuable. Beliefs and stories about the sky are also universal amongst the world's religions.[16] The particular meaning an individual or community may accord what is seen in the sky is, as Clive Ruggles and Nicholas Saunders suggest, 'as much a cultural as an astronomical one'.[17] Aveni showed by considering the !Kung, and Muris hunter-gatherers of Africa, the Polynesian sailors of the Pacific, and the Pawnee and Inca people of the Americas that these disparate groups all had strong and varying relationships with the sky.[18] Similarly, Ruggles and Saunders, in an exploration of the ancient astronomies found in Africa, Asia and Central America, demonstrate the great variety of astronomical practices found in those areas.[19] From an archaeo-astronomical perspective, Edwin

14 Anthony Aveni, *Conversing with the Planets: How Science and Myth Invented the Cosmos* (Boulder, co: University Press of Colorado, 2002), p. xiii.

15 John D. Barrow, *The Artful Universe* (Oxford: Clarendon Press, 1995), pp. 142–43.

16 Campion, *Astrology and Cosmology*, p.1.

17 Clive Ruggles and Nicholas Saunders, 'The Study of Cultural Astronomy', in *Astronomies and Cultures*, eds. Clive Ruggles and Nicholas Saunders (Boulder, co: University Press of Colorado, 1993), p.1.

18 Anthony Aveni, *People and the Sky: Our Ancestors and the Cosmos* (London: Thames and Hudson, 2008).

19 Ruggles and Saunders, *Astronomies and Cultures*.

C. Krupp's study focuses on various alignments found in megalithic sites in Europe, the Near East, and North and South America, and explores how observing the sky played a role in the cultural evolution of those peoples.[20]

Sometimes sky stories may be shared across different cultures separated by millennia. Consider the depiction of the constellation of Taurus the Bull. It is seen in both the Palaeolithic cave paintings at Lascaux, France and later in the mythology of the Babylonians.[21] The Pleiades however is probably the most universally significant star cluster and stories associated with these seven stars pop up in many cultures. In Norse mythology they are Freyja's hens, for the Native American Blackfoot tribe they are orphans and in Indonesia, seven princesses. Another example of cross-cultural sky stories is the 'ladder to the sky', a motif found in sources as diverse as the Old Testament Book of Genesis where it appears to the patriarch Jacob in a dream, amongst the indigenous Milingimbi people in Australia where the ladder is made from fish vertebrae and as a gigantic beanstalk in the English fairy tale, Jack and the Beanstalk.

However, without personal experience of observing certain sky features, some cultural references and stories may be missed or not understood. Slovick considers that to understand the references in an American folksong written by fugitive slaves in the nineteenth century – 'de river

20 Edwin C. Krupp, *Echoes of the Ancient Skies: The Astronomy of Lost Civilizations* (New York: Dover Publications, 2003).

21 Nicholas Campion, *A History of Western Astrology. Volume 1: The Ancient and Classical Worlds* (London: Continuum Books, 2008), p.14.

ends atween two hills, follow de drinkin' gou'd' – one needs to know that at that time the Big Dipper was likened to a drinking gourd and that for those slaves going north meant freedom.[22] A more recent example, in this case of the enduring cultural significance of Polaris the North Star, is, 'A Simple Response to an Elemental Message', a collaboration between the University of Edinburgh, the Royal Observatory of Edinburgh, and the UK Astronomical Technology Centre along with other partners.[23] Polaris was chosen because in the northern hemisphere many people will have seen it, it is easily identifiable and has long been used by stargazers and travellers as a reference point. In autumn 2016, messages from the public in response to the question: 'how will our present environmental interactions shape the future?' will be converted into radio waves and broadcast towards Polaris, taking 434 years to get to their destination. The responses will then be used to find out if the concerns people have about the environment are similar across the world.

Experiences of the sky are also recorded in the form of personal testimony, and there is a growing body of literature describing individuals' appreciation of the night sky. Much of this recent testimony focuses on the emotional impact of stargazing, for example Bogard's own stories, and those of others he has collected, many of which mourn

22 Slovick, 'Towards an Appreciation'; John A. Lomax and Alan Lomax, *American Ballads and Folk Songs* (New York: Dover Publications, 1994), p. 228.

23 University of Edinburgh 'Artistic odyssey to send messages to stars', http://www.ed.ac.uk/news/2016/starmessage-030216 [Accessed 12 Jan 2015].

the loss of 'real night' due to increasing light pollution.[24] Bogard himself grew up in a well-lit suburb in the US although spent summers near a lake where skies were much darker. Whilst Michael P. Branch insists that the increasing loss of visible stars in the Pleiades is linked to the loss of animal species, Christina Robertson identifies artificial lighting as the cause of the loss of certain species of bird.[25] Such testimonies celebrating dark skies may be seen as 'an urgent call to action', offering another source of backing for the Dark Sky movement.[26]

Furthermore, the growing Dark Sky movement, with the IDA as its main proponent, can be considered as fostering the human desire to see the night sky in practical ways through raising awareness on light pollution and developing a conservation programme for Dark Sky places.[27] The movement thus both sits within and is a development of the 'preservationism' stream of the environmentalism movement which, as Robyn Eckersley notes, encourages conservation of large areas of wilderness.[28] The 'Core' and 'Buffer' zone structure used in dark sky places is modelled on the design of UNESCO Biosphere Reserves. Literature on the Dark Sky movement itself is in its infancy as the

24 Bogard, *Let There Be Night*, p. 2; Bogard, *End of Night*.

25 Michael P. Branch, 'Ladder to the Pleiades', in Bogard, *Let There Be Night*, p. 83; Christina Robertson, 'Circadian heart', in Bogard, *Let There Be Night*, pp. 175–76.

26 Bogard, *Let There Be Night*, p. 6.

27 IDA, 'About IDA'.

28 Robyn Eckersley, *Environmentalism and Political Theory: Towards an Ecocentric Approach* (London: UCL Press Limited, 1992), p. 39.

movement only commenced in 1988.[29] Studies conducted thus far by the movement, or on its behalf, focus almost exclusively on the effects of light pollution on wildlife and humans and outdoor light design and control.[30] One study however, by Daniel Brown, conducted within the field of higher education on place-based learning, explores the possibility of increased sustainable behaviour regarding light pollution with the Peak District Dark Sky Group.[31] He concludes that this type of learning can lead to transformed behaviours.[32] Also, in 2015 Glasgow University and Forest Research (Scotland) announced a new collaborative doctoral project, 'Skies Above, Earth Below: Mapping the Values of the Galloway Forest Dark Sky Park'. The project seeks to investigate how local communities identify with and celebrate the dark sky in Galloway, the impact of the Park on local land use and 'the return of the dark as a contemporary, creative condition of living in the region'.[33]

The benefits to humankind of preserving dark skies is referred to in the objectives of the Starlight Initiative which encourages, 'defence of the values associated with

29 IDA, 'About IDA'.

30 *Ecological Consequences of Artificial Night Lighting*, eds. C. Rich and T. Longcore (Washington, DC: Island Press, 2006).

31 Daniel Brown, *How Can Higher Education Support Education for Sustainable Development? What Can Critical Place-Based Learning Offer?* [Unpublished MS, University of Nottingham, 2013].

32 Brown, *Higher Education*, p. 68.

33 University of Glasgow, 'Postgraduate Research Opportunities', http://www.gla.ac.uk/schools/ges/research/postgraduate/#/ahrccollaborativestudentship:'skiesaboveearthbelow' [Accessed 30 Mar 2016].

the night sky and the general right to observe the stars'.[34] In other words, a night sky free from pollution is considered to be a basic human right, comparable to other socio-cultural and environmental rights. The Declaration was adopted on the occasion of the Starlight Conference held in La Palma in 2007. The Initiative however provides no information as to what those values might be other than the 'material and intangible cultural heritage associated with astronomy'.[35] Similarly the Starlight Initiative states a sky free of light pollution has, 'an impact on the development of all peoples', but it does not specify what that impact might be.[36]

COMMERCIALISATION: ASTRONOMICAL TOURISM

Astronomical tourism, or what is now more commonly called astro-tourism, is by no means a modern day phenomena. People have travelled for centuries to sites such as Callanish on the Isle of Lewis in Scotland and the Bighorn Medicine Wheel in Wyoming for a number of reasons, including the celebration of particular astronomical events

34 The Starlight Initiative, 'Objectives of the Starlight Initiative', http://www.starlight2007.net/index.php?option=com_content&view=article&id=199&Itemid=81&lang=en [Accessed 23 Nov 2013].

35 The Starlight Initiative, 'Objectives of the Starlight Initiative'.

36 *StarLight: Declaration in Defence of the Night Sky and the Right to Starlight (La Palma Declaration)*, International conference in defence of the quality of the night sky and the right to observe the stars, La Palma, Canary Islands, Spain, (April 19–20 2007), eds. Cipriano Marín and Jafar Jafari (La Palma: Starlight Initiative, 2007), p. 3.

such as solstices but also to observe eclipses and auroras.[37] Some latter-day visitors still travel with these same aims – witness the large numbers who visit Stonehenge each year at the summer and winter solstices. However astro-tourism has now developed to include visiting particular sites with the specific intent of appreciating their unpolluted night skies, often with the additional intention of gaining knowledge for astronomical, cultural or environmental activities. Tourist activities such as these may be considered to come under the umbrella of sustainable tourism, that is, tourism which aims to ensure that a resource's, 'economic, social and aesthetic needs can be fulfilled while maintaining cultural integrity, essential ecological processes, biological diversity and life support systems'.[38]

Astro-tourism also comes in different guises. In May 2015 the Royal Astronomical Society (RAS) awarded £500,000 to six different projects in an effort to widen participation in astronomy and geophysics.[39] One project, 'Stepping Out: Astronomy Short Breaks for Carers', organised by Care for Carers, Our Dynamic Earth, the Royal Observatory Ed-

37 Glyn Daniel, 'The Forgotten Milestones and Blind Alleys of the Past', *RAIN* (Royal Anthropological Institute of Great Britain and Northern Ireland) 33 (1979): pp. 3–6, p.4; R. A. Williamson, *Living the Sky: The Cosmos of the American Indian* (Norman, OK: University of Oklahoma Press, 1984), p. 2

38 Institute for Tourism, 'What is sustainable tourism?' http://www.iztzg.hr/en/odrzivi_razvoj/sustainable_tourism/ [Accessed 12 Feb 2016].

39 Royal Astronomical Society, '£500k for Public Engagement in Astronomy and Geophysics: Six Teams Win RAS Funding', http://www.ras.org.uk/news-and-press/news-archive/259-news-2015/2629-460k-for-public-engagement-in-astronomy-and-geophysics-six-teams-win-ras-funding [Accessed 7 Mar 2016].

inburgh and Cosmos Planetarium, offers unpaid family carers a few days away from their caring responsibilities on the island of Coll, a Dark Sky Community in the Inner Hebrides off the west coast of Scotland. The aim is to help carers discover or rekindle an interest in astronomy whilst at the same time alleviating the stress of coping with challenging caring situations although the project is not specifically aiming to improve carers' wellbeing. When I contacted Robert Massey, RAS's Deputy Executive Director, to ask if the RAS were considering looking further into the effects on wellbeing of skywatching he acknowledged that there were no current plans to do so and that this was an area that has not had much exposure until recently.[40]

Chaco Canyon in New Mexico is a destination that has attracted astro-tourists for centuries; many of the monumental buildings built by the Chacoan people are aligned with solar and lunar cycles.[41] Nowadays, the Canyon's dark sky is recognised as a resource in itself and the Chaco Culture National Historical Park houses a visitor centre and a thriving night sky programme.[42] The US National Park Service defines areas where there is little or no light pollution at night as 'natural lightscapes', and much work goes on into protecting this endangered resource for both

40 Author's personal correspondence with Robert Massey 23 Mar 2016.

41 National Park Service, 'Chaco Culture National Historical Park New Mexico', https://www.nps.gov/chcu/index.htm [Accessed 19 Feb 2016].

42 National Park Service, 'Chaco Canyon'.

humans and nocturnal animals.[43] Moreover, lightscapes can also be both a cultural resource, 'part of the historical fabric of a place', as is the case with Chaco Canyon, and an economic resource, in areas such as the Colorado Plateau in the southwestern US.[44]

Returning now to Sark, in the IDA application Sark Tourism has 'long recognised the draw of true night-time darkness to visitors from mainland towns and cities' and hopes to further promote astro-tourism in order to bring increased economic benefit to the island. According to Eduardo Fayos-Solá *et al.*, the destinations best suited to astro-tourism have, 'very special characteristics, which makes for a likely favourable strategic positioning in domestic and even international markets'.[45] Sark's idiosyncratic system of government, its lack of outdoor lighting and car free roads and absence of many of the other trappings of modern day tourism suggest it meets these criteria. In addition, many inhabitants of the island were also coming round to seeing the benefits of labelling and marketing as a resource a phenomenon – Sark's very dark sky – which they had previously taken for granted. This may be compared to the marketing of other resources (which may at one time also

43 National Park Service, 'Natural Lightscape Factsheet', http://www.concessions.nps.gov/docs/concessioner%20tools/Natural_Lightscape_Factsheet.pdf [Accessed 23 Oct 2015].
44 National Park Service, 'Lightscape /Night Sky' https://www.nps.gov/ever/learn/nature/lightscape.htm [Accessed 23 Oct 2015].
45 Eduardo Fayos-Solá, Cipriano Marín, and Jafar Jafari, 'Astrotourism: No Requiem for Meaningful Travel', *PASOS: Revista de Turismo y Patrimonio Cultural* 12.4 (2014): pp. 663–71, p. 663.

have been 'undervalued') such as country parks, nature reserves, historic buildings and sites, by local authorities and cultural/natural heritage organisations such as English Heritage and Historic Scotland.

As regards Sark's sky being a cultural resource, there are a number of megalithic monuments, many with astronomical significance, on Sark and the other Channel Islands, which are likely to have been destinations for megalithic travellers. Sir Barry Cuncliffe, previously head of the Institute of Archaeology at the University of Oxford, has been conducting archaeological excavations of Sark's megaliths since 2009. His team's explorations have focused on the dolmen above Vermandaye Bay, and they have located a footprint of 'lost' standing stones near La Sablonnerie Hotel on Little Sark. Interestingly, a modern-day interpretation of a stone circle, Sark Henge, was erected on a cliff on the southeast coast of Sark near to Derrible Bay in August 2015 as part of the 450th anniversary celebrations. It consists of nine pink granite stones which were formerly gateposts with four of the stones aligned with the sun – on the summer solstice, the winter solstice, the morning sunrises and the spring/autumn equinox. The remaining five stones are aligned with landmarks – St. Ouen (Jersey), La Coupee (Sark), L'Etac (Jersey), Sark Mill (the highest point in the Channel Islands) and Alderney. It is fast becoming a new visitor attraction. Speaking to locals at the Henge the day I visited, I was amused by their speculations that in the distant future visitors might assume it was megalithic and construct all kinds of theories as to its purpose.

Increasingly there is evidence that across the globe astro-tourism is burgeoning, with associated economic benefits. In 2012, over 50,000 night-sky related visits to Bryce Canyon National Park in the Colorado Plateau contributed over $2 million to the local economy.[46] The European Sky Route project which began in 2013, and is co-financed by the European Union, aims to attract new visitors by developing a tourist route across five countries that links sites of astronomical interest with the primary aim being, 'the development of memorable star tourism experiences'.[47] The participating countries have begun work on projects that include combining astronomy with mythology (Greece), ancient history and archaeological resources (Bulgaria) and rural tourism and nature (Poland and Spain).[48] Further, less developed areas, for example Mbozi in southern Tanzania (which boasts the sixth largest meteorite in the world) and Taxiarchis, a small mountain village in Greece, are viewing these new opportunities with interest and beginning to lay their own plans for increasing visitors. It has also been suggested that there are economic advantages in protecting night skies owing to the decrease in light pollution. More judicious use of outdoor lighting can lead to

46 National Parks Conservation Association, 'Destination Darkness', www.npca.org/articles/341-destination-darkness [Accessed 12 Jan 2016].

47 EU Sky Route, 'European Astro Tourism Route', http://www.euskyroute.eu/european-astrotourism-route/ [Accessed 12 Dec 2015].

48 EU Sky Route, 'EU Sky Route Newsletter', http://www.euskyroute.eu/wp-content/uploads/2015/04/eu-sky-route-newsletter-160315_site.pdf [Accessed 12 Dec 2015].

large financial savings, and for the US this has been estimated to be as great as $2 billion each year.[49]

In particular, the areas which have achieved dark sky status, are now beginning to reap measurable economic rewards. A survey of thirty-five businesses in Galloway, including guest houses, bed and breakfasts, hotels and self-catering properties, reported 77% of them seeing an increase on the number of bed nights as a result of the Dark Sky Park (2012).[50] Similarly, in Northumberland 50% of businesses attending tourism business workshops reported dark sky tourism was having a positive impact on their bookings and many businesses have gone on to organise stargazing breaks.[51] The workshops were part of the, 'Animating Dark Skies Partnership' which was funded by the Department of Environment, Food, and Rural Affairs. The intention was to help foster the stargazing tourism sector and gave practical information regarding steps to take to boost astro-tourism, for example, encouraging the use of environmentally friendly lighting, providing guests with basic equipment, signposting to astronomy apps and running events. As a result, guests at local hotels are now routinely provided with night-vision torches, and deckchairs are put out at night. Hoteliers with some knowledge of as-

49 Tim Hunter and David Crawford, 'The Economics of Light Pollution', ASP *Conference Series* 7 (1991): p. 99.

50 North Penines, 'Animating Dark Skies Project Partnership', http://www.northpennines.org.uk/Lists/DocumentLibrary/Attachments/508//RichardDarnNPennTourismSeminar2014StargazingPresentation.pdf [Accessed 12 Dec 2015].

51 North Penines, 'Animating Dark Skies'.

tronomy can also apply for a badge confirming that their hotels are 'Dark Sky Friendly'.

In many dark sky places, having an observatory accessible to the public is increasingly being seen as not only a useful but necessary addition to attract more astro-tourists. In addition, an increasing number of observatories are appearing or in the process of being developed, across the world. The Scottish Dark Sky Observatory (SDSO), located on the edge of Galloway Forest Dark Sky Park, opened in 2012. It also has its own mobile planetarium which is used to take a planetarium experience out to the public. In March 2015 the Park advertised for four Dark Sky Rangers to enable further development of stargazing events. Meanwhile over the Scottish border, the Kielder Observatory in Northumberland Dark Sky Park (which already attracts around 24,000 visitors annually) has plans to build an 'astrovillage', which will include the biggest public observatory in the world, accommodation for visiting astronomers, a sixty-seat planetarium and a specially built £500,000 telescope with a one metre wide aperture. The cost for the project is estimated to be around £8.5 million and the hope is to increase visitor numbers to 75–100,000 a year. In Ireland, Kerry International Dark Sky Reserve recently got permission to build a small observatory, while Zselic Dark Sky Park in Hungary have just opened a $3.3 million visitor facility which includes a public observatory and planetarium. Meanwhile in October 2015 Sark unveiled its own observatory, albeit on a more modest scale. Recognising the impact this might have, Daschinger commented, 'I think we are beginning to get more and more professional astronomers and now we have the observatory, that's very much better for them'.[52]

During my visit to Sark in September 2015, four years after the award of dark sky status, I was curious as to what evidence I might find regarding the promotion of astro-tourism. The 2016–2017 Sark Tourism brochure devotes a full page to 'Sark at Night', and between July and August 2015 Sark Visitor Centre had a 'Dark Skies' exhibition. The SAstroS *Starfest* events run regularly, and the partnership between SAstroS and Stocks Hotel on Sark is now well established. On the La Vallette campsite website, they advertise 'at night it is the perfect place to lie on the dewy grass to admire the various constellations and appreciate that in Sark there is no pollution to mar the view of the stars in the sky'.[53] This illustrates a clear positioning by the campsite owners of La Vallette as a destination to watch the night sky. I also noticed the Vieux Clos Guest House offering guests the use of a telescope and binoculars. Rosie told me someone had commented in her bed and breakfast visitors' book that Sark was a wonderful place to see the stars, although she does not explicitly mention the dark sky in her own marketing. I asked at the Visitor Centre how many inquiries were made about the dark sky and was told there was a 'steady stream' of emails from people planning to visit. Also, when visitors dropped into the Centre for oth-

52 BBC News, 'Sark's Astronomical Observatory Opens', http://www.bbc.co.uk/news/world-europe-guernsey-34495607 [Accessed 12 Dec 2015].

53 Sercq – La Valette Campsite, http://www.sercq.com/la_Valette_Campsite.html [Accessed 5 May 2016].

er information seeing the photographic display on Sark's night sky behind the reception desk often sparked further questions. It is unclear as yet how SAstroS in conjunction with Sark Tourism can best market their particular unique selling point, the observatory, and ensure easy access to visitors.

HERITAGE AND NOSTALGIA TOURISM

Heritage tourism, that is tourism that engages with a particular location's cultural tradition, is increasingly big business. According to Historic England, heritage tourism accounted for 2% of the UK's gross domestic profit (GDP) in 2011.[54] World Heritage Sites alone bring in £61.1 million each year with associated benefits to local economies.[55] Within the field of heritage tourism, the sub-field of 'nostalgia tourism' is receiving increasing attention.[56] The concept of 'nostalgia' appears to have existed before the word itself,

54 Historic England, 'Heritage and the Economy', p. 1, https://content.historicengland.org.uk/images-books/publications/heritage-and-the-economy/heritage-and-the-economy-2015.pdf [Accessed 12 Dec 2015].

55 UNESCO, 'Wider value of UNESCO to the UK', http://www.unesco.org.uk/wp-content/uploads/2015/05/Wider-Value-of-UNESCO-to-UK-2012-13-full-report.pdf [Accessed 12 Dec 2015].

56 F. Davis, *Yearning for Yesterday: A Sociology of Nostalgia* (New York: Free Press, 1979); M. Kibby, 'Tourists on the Mother Road and the Information Superhighway', in *Reflection on International Tourism: Expressions of Culture, Identity, and Meaning in Tourism*, eds. M. Robinson, P. Long, N. Evans, R. Sharpley and J. Swarbrooke (Newcastle: University of Northumbria), pp. 139–49.

featuring strongly in Homer's *Odyssey*, and having different meanings attached to it at different times. The word derives from the Greek words '*nostos*', meaning return to one's land of birth, and '*algos*', meaning suffering or pain. It was first described in the seventeenth century by Johannes Hofer who, while acknowledging it was a yearning to return home, viewed it as a type of neurological disease.[57] More recently, although some continue to see nostalgia as mainly negative in the sense that it involves grieving for a past that is irretrievably lost, others suggests it can be seen as a more positive emotion.[58]

In the sphere of nostalgia tourism however, nostalgia is generally understood to be the contrasting of certain interpretations of the past with those of the present and in this process, the past becomes viewed through 'rose-tinted glasses' and thus associated with positive qualities and emotions such as, 'beauty, pleasure, joy, (and) satisfaction', while correspondingly the present is viewed as, 'more bleak, grim, wretched, ugly, deprivational, unfulfilling, (and) frightening'.[59] 'Nostalgia tourists' are usually seen as falling into two categories, 'real' and 'historical', with 'real' meaning those tourists who are revisiting an actual

57 J. Hofer, 'Medical Dissertation on Nostalgia', trans., C. K. Anspach, *Bulletin of the History of Medicine*, vol. 2, pp. 376–91.

58 D. G. Hertz, 'Trauma and Nostalgia: New Aspects of the Coping of Aging Holocaust Survivors', *The Israel Journal of Psychiatry and Related Sciences* 27.4 (1990): pp. 189–98; Davis, *Yearning*; M. B. Holbrook, 'Nostalgia and Consumption Preferences: Some Emerging Patterns of Consumer Tastes', *Journal of Consumer Research* 20.2 (1993): pp. 245–56.

59 Davis, *Yearning*, pp.14–15.

experience they have had themselves, and 'historical' being those tourists who seek to visit an idealised environment they have not had prior direct experience of.[60] Moreover, research shows that marketing which appeals to the nostalgia customer's 'empathy and idealization of self' greatly increases the selling power of the product.[61] With both types of nostalgia tourists therefore there is a bittersweet longing for the past and often this longing appears to be fuelled by some kind of anxiety, fear or discontent over their current circumstances.[62] Apparently we are more likely to feel nostalgic for childhood and adolescent experiences, and in the US there has been a particular focus in marketing the types of objects and images which trigger memories of these times such as 'retro' sweets, toys, books.[63]

Whilst the Channel Islands have long been a popular tourist destination, the islands are not usually branded together for the purposes of tourism. Whilst this inevitably leads to a level of competition between them, in many ways this can be a benefit as it enables each island to focus on its own island identity. Sark for example trades particu-

60 Dale W. Russell, 'Nostalgic Tourism', *Journal of Travel and Tourism Marketing* 25.2 (2008): pp. 103–16, p. 104.

61 Barbara Stern, 'Historical and Personal Nostalgia in Advertising Text: The *Fin de siècle* Effect', *Journal of Advertising* 21.4 (1992): pp. 11–22, p. 11.

62 G. M. S. Dann, 'Tourism: The Nostalgia Industry of the Future' in *Global Tourism: The Next Decade*, ed W. F. Theobald. (Oxford: Butterworth-Heinemann Ltd., 1994), pp. 55–67, p. 65; R. Hewison, *The Heritage Industry: Britain in a Climate of Decline* (London: Methuen London Ltd., 1987), p. 45.

63 M. B. Holbrook, 'Nostalgia Proneness and Consumer Tastes', in J. A. Howard, ed., *Buyer Behavior in Marketing Strategy*, 2nd ed., (Englewood Cliffs, NJ: Prentice-Hall, 1994), pp. 348–64.

larly on the lack of street lighting and cars. Sark Tourism's strapline, 'Sark, a world apart', and a lot of its marketing material harken back to a pre-industrial era. The Sark 'experience' is likened to 'the stuff of Famous Five and Swallow and Amazons. Wonderful memories are made here, by adults and children alike'. A lot of the events running throughout the tourist season are family-friendly, such as treasure trails, boat trips, rock pool rambles and church fetes. Good Friday sees crowds gather at Beauregard pond for the traditional sailing of model boats. In other words, visitors are being consciously invited to recreate a rural (1950s?) childhood (real or imagined) for themselves and, if they have any, for their own children too. Visitors can get off the ferry and climb onto the 'toast rack', a cart hooked up to the back of a tractor, to take them up Harbour Hill to collect bikes, walk to their accommodation or take a horse-drawn carriage. The world outside Sark with its 'dangerous' roads, crime, strangers and light pollution can be left behind. This intentional evoking of nostalgia in visitors (of all ages) leads to the assumption that some visitors can therefore be considered 'nostalgia tourists'. Although it is important of course to remember that others will visit the island purely for recreational or other purposes.

Whilst Sark's economy is heavily reliant on tourism, its tourism infrastructure is minimal. This results in certain events, such as the annual three-day Sark Folk Festival, necessarily restricting the amount of tickets available. A lack of beds on the island means organisers cannot increase capacity beyond about 1,400. This usually results in tickets selling out very quickly (for the 2016 event they were snapped up within half an hour of going on sale). Limiting numbers also has the benefit of guarding against over-commercialisation and inevitably results in a level of exclusivity attaching to such events.

Although most visitors to Sark would not expect nightclubs, twenty-four-hour shopping malls, drive-in burger bars, and so forth, certain tourist facilities are however required. There are two independently run hotels on the island, ten guest houses, sixteen self-catering properties and two campsites. In addition, there are a substantial number of businesses (including a further four hotels), which are owned by Sir David and Sir Frederick Barclay. The brothers also own Sark Estate Management (SEM), which has several vineyards and operates Sark Island Hotels. Guests at SEM hotels can go on guided tours of Brecqhou. Since the Barclays bought Brecqhou in 1993 there have been tensions and legal challenges between them and Sark's parliament, including the contesting of Sark's laws and system of government and Sark's jurisdiction over Brecquou. In 2008, the brothers closed their businesses for almost two weeks in protest at the failure of the majority of the candidates they had supported to win seats in Sark's first fully democratic parliamentary elections. In 2015, despite tourism figures being healthy, they again closed their hotels, citing as their justification the lack of a customs post on Sark denying local businesses access to the French tourism market. Later that year SEM announced that the closures would continue into 2016. This was despite overall visitor numbers to Sark rising in 2014 as compared to the previous year.[64] These closures have led to a loss of jobs and, unsurprisingly, resulted in a complex situation that has yet to be resolved.

In 2012 a Sark Island-Wide Opinion Survey carried out by the Chief Pleas, which included a SWOT (Strengths, Weaknesses, Opportunities and Threats) Analysis, identified that there was further potential to develop more specialist tourism, for example the promotion of dark skies, links to cruise liners, short breaks and activity based tourism.[65] Against this backdrop, there appears to be particular interest amongst residents and local businesses in promoting tourism to Sark seeking ways of tapping into the astrotourism and nostalgia tourism markets and thereby adding value to the tourist experience. An increasing number of local people are also moving into offering bed and breakfast accommodation in their own homes. This is another example of the island's long tradition of creatively managing to preserve the best aspects of their way of life when change is imposed from elsewhere. In addition, for Sark, as with many other small islands, creeping globalisation increasingly threatens insular ways of life.

64 BBC News, 'Barclay brothers' Sark hotels to close', http://www.bbc.co.uk/news/world-europe-guernsey-30035969 [Accessed 1 Apr 2016]

65 Sark Chief Pleas, 'Sark Island-Wide Opinion Survey SWOT Analysis', http://www.gov.sark.gg/Downloads/Reports/SWOT_Analysis_290912.pdf [Accessed 1 Apr 2016]

FEAR OF THE DARK

Moving on from concerns and opportunities of nostalgia tourism to another facet of the work: the fear of the dark. Fear of darkness seems embedded in our psyche and a dark night is often seen as dangerous; in its most severe form this fear is known as nyctophobia. Edmund Burke considered it to be, 'universally terrible in all times and in all countries', and A. Roger Ekirch called it, 'our oldest and most haunting terror'.[66] Martin Seligman goes further and suggests that although the dangerous animals, neighbouring tribes, situations, etc., that threatened our ancestors no longer exist for most modern humans, we are still 'biologically prepared' to fear these things.[67] Tim Edensor however does not believe that it was necessarily always the case and certainly not so in all cultures.[68] It is recognised as one of the most common fears amongst children and the vast quantity of resources available to help them overcome their anxiety is testament to this. Many of us probably recall childhood fears of monsters under the bed and asking a parent to check they had gone before we could safely get to sleep. Ironically, one popular contemporary method of

66 Edmund Burke, *On the sublime and beautiful*, https://ebooks.adelaide.edu.au/b/burke/edmund/sublime/part4.html [Accessed 1 Apr 2016]; Ekirch, *At Day's Close*, p. 3.

67 Martin E. P. Seligman, 'Phobias and Preparedness', *Behavioral Therapy* 2.3 (1971): 307–20.

68 Tim Edensor, 'The Gloomy City: Rethinking the Relationship between Light and Dark', http://usj.sagepub.com/content/early/2013/09/24/0042098013504009.full [accessed Mar 14 2014].

attempting to calm a child's fear is to pepper their bedroom ceiling with stick-on luminous stars – whereas actually taking the child out into a starry night would probably not even be considered by most parents! Although it is generally assumed children will eventually grow out of their fear, in fact many do not, and go on to become adults who remain afraid of the dark.

The roots of this terror can perhaps, as Ekirch speculates, be traced back to our prehistoric ancestors who felt profound fear 'amid the gathering darkness and cold' because they could not be sure the sun would return – but they also had to contend with many very real nocturnal dangers, such as dangerous animals.[69] Before gas lighting was introduced in the UK and France in the eighteenth century and the first appearance of Thomas Edison's electric bulbs in a New York street in 1879, there were good reasons to tread cautiously after nightfall, as in most urban areas there would be rubbish and excrement lying around. In rural areas more usual hazards of the night would be 'fallen trees, thick underbrush, steep hillsides and open trenches'.[70] A few years ago my friend Alex sustained a serious injury when cycling at night down a steep hill; he had forgotten his head torch, did not see a large stone and ended up in a ditch. As Sark has no acute medical services he had to be taken by ferry to Guernsey and treated at the Princess Elizabeth Hospital. Clearly accidents still await the unwary traveller at night on Sark.

69 Ekirch, *At Day's Close*, p. 3.
70 Ekirch, *At Day's Close*, p. 123.

Although for many people, particularly those living in urban areas, these hazards no longer exist to anything like the same degree, responses to darkness can be variable. Whilst some (including myself) relish being out in darkness and find it alluring, others feel frightened and unnerved by it. This was brought home to me recently when leaving a public lecture on astronomy at the Royal Observatory in Edinburgh, I overheard several people voice their concerns about walking out into the darkness. I had assumed (wrongly as it turned out) that those attending such a lecture would be comfortable with being out on dark nights. My father also recently reminded me that for those people, like him, who as a child had lived through the enforced blackouts in Glasgow during the Second World War darkness may be associated with the fear of German bombings. Whilst the younger residents on his street tolerated broken street lighting, he complained to his local council, as he, 'didn't like going out then, it was like the blackout'. However the membership of the BAA increased during the Second World War, perhaps because people had greater access to the night sky.[71] It cannot be assumed therefore that darkness is universally appreciated and indeed both darkness and illumination may, as Edensor suggests be, 'loaded with contested values' as, 'what is for some a scene of safety and, cheeriness, might for astronomers testify to the dilution, even disappearance of the nocturnal celestial sky'.[72]

71 R. McKinn, *The History of the BAA: The First Fifty Years* (London: BAA, 1990).
72 Edensor, *Gloomy City*.

Darkness often appears to have been viewed, not only as something to be feared, but something potentially evil. The symbolic role darkness played, particularly in medieval religious societies may represent, according to J. Galinier *et al.*, 'pagan obscurantism - deviancy, monstrosity, diabolism'.[73] Robert Shaw notes, 'the night was a time when demons, criminals and others who presented a threat were imagined to be present in the landscape'.[74] Witches and goblins may have abounded but as Craig Koslovsky reminds us darkness (and more broadly night-time itself) can also be full of contradictions, 'both devilish and divine, restful and restive, both disciplined and ungovernable'.[75] Similarly Luc Gwiazdzinski characterises night as a time when the normal rules of behaviour may be ignored and creativity can flourish.[76] In contrast, C. G. Jung believed Western civilisation, particularly, should pay attention to its need for connection to its 'unconscious wildness' as the gradual conquering of external 'wild nature' has led to a taming of our own 'inner wildness'. He saw evidence of this in his consulting room in patients' common fear of

73 J. Galinier, A. Becquelin, G, Bordin, *et al.*, 'Anthropology of the Night: Cross-disciplinary Investigations'. *Current Anthropology* 51.6 (2010): pp. 819–47, p. 820.

74 Robert Shaw, 'Controlling Darkness: Self, Dark and the Domestic Night', *Cultural Geographies* 22.4 (2015): pp. 585–600, p. 589.

75 Craig Koslovsky, *Evening's Empire: A History of the Night* (Cambridge: Cambridge University Press, 2011), p. 5.

76 Luc Gwiazdzinski, 'La nuit dernière frontière/Night – The Last Frontier', *Les Annales de la Recherche Urbaine, Plan Urbanisme – Construction – Architecture* 87 (2000): pp. 81–89.

the 'outer wildness', wilderness areas, certain animals, and darkness.[77]

Through the centuries artificial illumination – whether by candles, lanterns or gaslight – helped allay fears about the 'things that go bump in the night', and, by now, people will generally resist giving up nocturnal lighting they have become accustomed to. This is likely to be because there has always been the assumption that people are somehow more ungovernable at night and more crime occurs then. Ekirch gives many examples to suggest that, in early modern Europe at least, this appear to have been the case.[78] The advent of street lighting in Paris was not welcomed by the certain groups in society who had been used to, for whatever reason, being out at night. In a sense night has been gradually, 'opened up, expanded into and colonised'.[79] More recently, Kopel and Loatman caution against, 'excessively severe Dark Sky laws', suggesting some individuals may resist giving up particular types of lighting for fear of increased crime or loss of privacy.[80] The argument that dark skies could be detrimental to human wellbeing and safety is however disclaimed by the IDA, who contest that outdoor night-time lighting does not necessarily prevent

77 C. G. Jung, *The Structure and Dynamics of the Psyche*, in C. G. Jung, *The Collected Works* 8, 2nd ed., trans. R. F. C. Hull, (Princeton, NJ: Princeton University Press, 1969), ¶ 67.

78 Ekirch, *At Day's Close*, p. 33.

79 Robert Shaw, 'Night as Fragmenting Frontier: Understanding the Night that Remains in an Era of 24/7', *Geography Compass* 9.12 (2015): pp. 637–47, p. 637.

80 Kopel and Loatman, *Dark Sky*, p. 1.

crime although it may help reduce fear of crime.[81] Similarly, a report by Stephen Atkins *et al.* found increased street lighting had little or no effect on crime.[82] In addition, a later study by Rebecca Steinbach *et al.* suggested that as long as all risks were considered carefully, local authorities can reduce street lighting without there being a subsequent rise in road collisions or crime.[83] On a more populist level, a recent conversation on *Streetlife*, an online social network which aims to 'help people make the most of where they live by connecting and sharing with neighbours', concerned new street lighting in an area near my home in Edinburgh.[84] Several people assumed that the local council's proposals for less bright lighting would inevitably lead to increased crime. One individual wrote, 'Has anyone noticed the new lighting in Findlay Gardens area? It is total blackout? I'm all for cutting light pollution but at the expense of someone being mugged? Or helping burglars rob homes due to the dark cover afforded by these new lights?' Another, 'My elderly parents feel really concerned with the

81 International Dark-Sky Association, 'Lighting and Crime. Information Sheet No. 51', http://www.darksky.org/assets/documents/is051.pdf [Accessed 29 Nov 2013].

82 Stephen Atkins, Sohail Husain and Angele Storey, *The Influence of Street Lighting on Crime and Fear of Crime*, Crime Prevention Unit Paper No. 28 (London: Home Office, 1991).

83 Rebecca Steinbach, Chloe Perkins, Lisa Tompson *et al.*, 'The Effect of Reduced Street Lighting on Road Casualties and Crime in England and Wales: Controlled Interrupted Time Series Analysis', *Journal of Epidemiology and Community Health* 69.11 (2015): 1118–24.

84 Streetlife, 'New Street Lighting', https://www.streetlife.com/home/my-conversations/ [Accessed 23 Oct 2015].

difference in that their street and are very concerned about burglars / muggers etc. They no longer feel safe walking to / from the bus stop at the end of the street!'[85] My reaction to these comments was to post that there was lots of evidence to say reduced street lighting does not increase crime and also most crime happens during the brighter summer months. As several people 'liked' my comments, I assume they helped allay their fears but the general feeling seems to be that reduced lighting inevitably leads to more crime even if no one had actually experienced this themselves.

Darkness however has not only been feared but also welcomed by some. Turning now to the attributes of darkness, Ekrich argues that during the early modern era in cities, 'the darkness of night loosened the tethers of the visible world'; everyone was free to roam without the constraints of the day and social status was less immediately discernible.[86] In actual practice most people probably chose to stay at home preferring to be warm and safe. Edensor also comments darkness may not only be 'synonymous with superstition, murky thoughts and illicit behaviour but replete with generative potentialities and affective possibilities'.[87] He further suggests that returning to a darker environment might enable people to come together at night, 'the potentialities of darkness might foster progressive forms of conviviality, communication and imagination'.[88] This

85 Streetlife.
86 Ekrich, *At Day's Close*, p.152
87 Edensor, *Gloomy City*, p. 13.
88 Tim Edensor, 'Reconnecting with Darkness: Gloomy Landscapes, Lightless Places', *Social & Cultural Geography* 14.4 (2013): pp. 446–65.

sense of darkness as being positively transformative is echoed by Shepherd Bliss who sees darkness as an, 'underappreciated force for healing'.[89] Watching the moon's rays he describes disconnecting from his problems and placing himself, 'within the larger context of the earth's bounty'.[90] Both Edensor and Bliss appear to be advocating that contrary to what our primitive survival mechanisms may indicate, darkness can be appreciated and welcomed.[91] William Sheehan goes further and suggests fear of darkness, and for some the excitement that goes with that, may account for *noctcaelador*, perhaps because there is a desire to reproduce the 'thrill'.[92] Also, the opportunity that darkness gives of cultivating closer connections with the natural world is suggested by Kathleen Dean Moore when she says, 'when stars blink on [...] then the structure of the built world begins to vanish'.[93]

THE 'NATURE' OF NATURE

Although it is a common human practice to observe the night sky and assign meanings and stories to this activ-

89 Shepherd Bliss, 'In Praise of Sweet Darkness', in Linda Buzzell and Craig Chalquist, eds., *Ecotherapy: Healing with Nature in Mind* (San Francisco: Sierra Club Books, 2009), p. 174.
90 Bliss, 'In Praise of Sweet Darkness', p. 182.
91 Edensor, *Gloomy City*, p. 13; Bliss, 'In Praise of Sweet Darkness', p. 174.
92 William Sheehan, *A Passion for the Planets: Envisioning Other Worlds, from the Pleistocene to the Age of the Telescope* (New York: Springer, 2010), p. 47.
93 Kathleen Dean Moore, 'The Gifts of Darkness', in Bogard, *Let There Be Night*, p. 12.

ity, what we observe going on in the heavens is often considered to be separate from earth and the natural world. Definitions of nature do not make specific mention of celestial bodies and sky features, which suggests that the sky is not generally considered to be part of nature. Tim Ingold points out that theories as to how people perceive the world, 'generally work from the assumption that this world is terrestrial', and he suggests the sky is indeed part of the landscape.[94] This view is echoed by Edensor when he speaks of how our notions of what we consider to be landscape usually omit the celestial and focus instead on, 'that which is of the earth'.[95] The Dark Sky movement itself is considered by some to have been formative in promoting, 'an understanding of landscape that incorporates aspects of cosmology, as well as generational and geological time frames'.[96] Elsewhere Ingold suggests we usually experience ourselves as separate from, rather than part of, this world and proposes that the landscape is a part of us and furthermore that we can have a 'felt' experience of this connection. Bogard takes this further, describing the intimacy of this connection for him between earth and sky: 'the moon climbs slowly, sometimes so dusted with rust and rose, brown, and gold tones that it nearly drips earth colours and seems intimately braided with Earth, it feels close, part of this world, a friend'.[97] When Edensor visited

94 Tim Ingold, *Being Alive: Essays on Movement, Knowledge and Description* (London and New York: Routledge, 2011), pp. 126–27.

95 Edensor, 'Reconnecting', p. 453.

96 Dunnett, *Contested Landscapes*, p. 14.

97 Bogard, *End of Night*, p. 29.

Galloway Forest Dark Sky Park at night, he was struck by how usually the horizon acts as a marker between the earth and sky, but on that particular night the sky was very dark and he found his attention drawn primarily to the sky.[98] This observation chimed with my own the first time I was out on a moonless night on Sark, and I tripped over because I lost sense of where the ground was! He also noted the difference in how close or far away we may experience the landscape, depending on whether it is observed during daylight, or at night when there is a lack of discernible detail and the boundary between earth and sky appears as a, 'largely undifferentiated realm that thwarts the usual sense that the landscape broadens out from the observer'.[99] Similarly, Duncan Wise, from the Northumberland National Park Authority comments, 'We tend to look at landscape as everything up to the horizon but what about what's above it?'[100]

NATURE AND WELLBEING

Although many studies explore the relationship between experiences in nature and positive outcomes for health and wellbeing, they do not usually include references to the sky.

98 Edensor, *Reconnecting*, p. 455.

99 Edensor, *Reconnecting*, p. 455.

100 *The Guardian*, 'Bright Future for "Dark Sky" Sites as Astrotourism Grows in Appeal', http://www.theguardian.com/science/2015/apr/12/dark-sky-tourism-northumberland-kielder-observatory-northern-lights [Accessed 12 Dec 2015].

Rachel Kaplan's research on the psychological benefits of looking out of a window at home forms a notable exception.[101] Although Kaplan considered whether viewing the sky and weather had any substantial psychological effect, the results were statistically insignificant when compared to viewing trees and landscapes.[102] However, her study appears to focus on viewing discrete 'pieces' of nature, whereas I would suggest it is difficult to imagine viewing the tree without also noticing the sky in the background!

Ecopsychology explores the idea that it is beneficial to human wellbeing to have a relationship with nature and this relationship may also have holistic, transformative and existential aspects to it. One key benefit is that such connections encourage ecological behaviour - we become more concerned with preserving the environment. This concept contrasts with what Deborah DuNann Winter describes as a, 'modern worldview that provides a set of beliefs that encourages us to use and abuse nature'.[103] Furthermore, Nick Totton suggests that for many people their relationship with nature is more powerful during childhood, and to rediscover this relationship as an adult is to, 'recover magic'.[104] There are many Walt Disney classic films for example which show children talking to plants and

101 Rachel Kaplan, 'The Nature of the View from Home: Psychological Benefits'. *Environment and Behavior* 33.4 (2001): 507–42.

102 R. Kaplan, 'Nature of the View', p. 535.

103 Deborah DuNann Winter, *Ecological Psychology: Healing the Split between Planet and Self* (New York: HarperCollins College Publishers, 1996), p. 29.

104 Nick Totton, 'The Practice of Wild Therapy', *Therapy Today* 25.5 (2014): pp. 14–17, p. 17.

animals – *The Jungle Book* comes to mind. Recent research by the Royal Society for the Protection of Birds (RSPB) explores and attempts to measure the connection children have with nature and suggests how to maintain this relationship throughout life in order to avoid what Richard Louv earlier described as 'nature deficit disorder', which he warned leads to, 'diminished use of the senses, attention difficulties, and higher rates of physical and emotional illnesses'.[105] Writing respectively in the late nineteenth and early twentieth centuries, William James and C. G. Jung may be seen as having contributed to the intellectual roots of ecopsychology, as both believed in the transformative and beneficial aspects of a human-nature connection for wellbeing although 'wellbeing' was not a term either of them actually used.[106] James promoted connecting with nature as a way of finding meaning and purpose in life, and he remarked, 'living in the open air and on the ground, the lopsided beam of the balance slowly rises to the level line; and the over-sensibilities and insensibilities even themselves out'.[107] Jung similarly remarked, 'we all need nourishment for our psyche. It is impossible to find such nourishment in

105 Royal Society for the Protection of Birds (RSPB), 'Connecting to Nature', http://www.rspb.org.uk/Images/connecting-with-nature_tcm9-354603.pdf [Accessed 17 Jul 2014]; Richard Louv, *Last Child in the Woods: Saving Our Children from Nature-Deficit Disorder* (Chapel Hill, NC: Algonquin Books of Chapel Hill, 2005), p. 34.

106 William James, 'On a Certain Blindness in Human Beings', http://books.google.com/books?isbn=0141956585 [accessed Dec 12 2013]; C. G. Jung, *Jung Speaking: Interviews and Encounters*, eds. William McGuire and R. F. C. Hull (Princeton, NJ: Princeton University Press, 1977).

107 William James, 'On a Certain Blindness'.

urban tenements without a patch of green or a blossoming tree'.[108] Yet even in the least likely places nature can still be found and a connection made. Working as a counsellor at a community health project in an area of multiple deprivation during the 1990s I saw many clients who spent their days in cramped Edinburgh tenements. To help tackle the area's social problems funding was found to plant a community woodland. For several of my clients helping plant the trees, maintaining the land and then relaxing there was wonderfully therapeutic and gave them a rare sense of purpose and achievement.

A number of authors, including David Abram, Freya Mathews and Roszak have gone on to explore in different ways the concept of a synergy between the wellbeing of human beings and the larger ecosystem of the natural environment.[109] While Abram stresses the importance of recognising a link between an individual's interior psychological world and the exterior world, questioning, after poet Rainer Maria Rilke, 'the inner' – 'what is it if not intensified sky'?[110] Mathews, however, critiques the atomistic cosmology whereby the world is seen as comprising many separate substances. She conjectures that this stance encourages disconnection between humans and their environment and, 'the right cosmology will dispose us towards

108 Jung, *Jung Speaking*, p. 202.

109 David Abram, *The Spell of the Sensuous: Perception and Language in the More-Than-Human World* (New York: Vintage Books, 1996); Mathews, *Ecological Self*; Roszak *et al.*, *Ecopsychology*.

110 Abram, *Spell*, p. 262.

a benign pattern of interaction with the environment'.[111] Mathews' view is echoed by Abram who sees phenomenology as a helpful approach for thinking about this disconnection, as phenomenology focuses on the qualitative dimension of experience and allows questioning of 'the modern assumption of a single, wholly determinable, objective reality'.[112] Roszak similarly berates the long-standing culture in the western world of the 'permissible repression of cosmic empathy' and the emotional, physical and spiritual distress which can result.[113] His claims of distress resulting from human-nature disconnection are however made on theoretical grounds with no empirical evidence to support the notion that a connection with nature is beneficial. This is in line with Jung's view that advancements in scientific understanding have had an adverse impact on humanity's contact with nature, 'and with it has gone the profound emotional energy that this symbolic connection supplied'.[114]

The fields of health and environmental psychology provide further evidence documenting the relationship between experiences in nature and positive outcomes for health and wellbeing. Judging by the considerable amount of literature, it can be argued that the concept of nature offering possibilities for psycho-physiological restorative effect is currently the most active research topic in the en-

111 Mathews, *Ecological Self*, p. 141.

112 Abram, *Spell*, p. 31.

113 Roszak, 'Where Psyche meets Gaia', in Roszak *et al.*, *Ecopsychology*, p. 10.

114 C. G. Jung, *Man and his Symbols* (New York: Random House, 1964), p. 95.

vironmental psychology field. Whilst 'wellbeing' is a problematic term to define, Rachel Dodge observes it is generally considered to encompass the emotional, physical and spiritual aspects of an individual's life.[115] There are two main schools of thought as regards what may be responsible for (any) beneficial effects of nature on wellbeing; these are S. Kaplan's Attention Restoration Theory and psychoevolutionary theory proposed by Ulrich.[116] These theories adopt conflicting viewpoints, with S. Kaplan concentrating on how nature can assist in the recovery from directed attention fatigue, and Ulrich on the positive emotional response and resulting reduction in stress, which may be experienced when looking at unthreatening nature.[117] T. Hartig and G. W. Evans however hold that it is possible to synthesise these two approaches, and S. Kaplan has now also taken this view forward.[118]

115 Rachel Dodge, Annette P. Daly, Jan Huyton and Lalage D. Sanderset, 'The Challenge of Defining Wellbeing'. *International Journal of Wellbeing* 2.3 (2012): pp. 222–35.

116 Stephen Kaplan, 'The Restorative Benefits of Nature: Towards an Integrative Framework', *Journal of Environmental Psychology* 16 (1995): pp. 169–82; Roger S. Ulrich, 'Aesthetic and Affective Response to Natural Environment', in I. Altman and J. Wohlwill, eds., *Behavior and the Natural Environment* (New York: Plenum Press, 1983).

117 S. Kaplan and J. F. Talbot, 'Psychological Benefits of a Wilderness Experience', in Altman and Wohlwill, *Behavior and the Natural Environment*; Ulrich, 'Aesthetic and Affective Response'.

118 T. Hartig and G. W. Evans, 'Psychological Foundations of Nature Experience', in T. Garling and R. G. Golledge, eds., *Behavior and Environment: Psychological and Geographical Approaches* (Amsterdam: Elsevier/North Holland, 1993), pp. 427–57.

Regardless of what may be responsible for nature's beneficial effects, research which supports the benefits include a study of recovery from mental fatigue and physical ill-health in different restorative environments by Hartig *et al.* which concluded natural environments had the greatest restorative effect.[119] The study included potentially hazardous environments such as wilderness and highlights that not all natural settings have to be safe to be restorative. R. Kaplan and S. Kaplan, in an overview of research on the importance of sensory stimulation in natural environments, note even the most mundane environment – such as the view from an office window – can potentially enhance wellbeing.[120] This begs the question as to whether nature has to be aesthetically pleasing to be restorative, an area Roger S. Ulrich explores in relation to aesthetic and affective responses in natural environments.[121] Furthermore, S. Kaplan's Attention Restoration Theory suggests one of the reasons nature experiences can be restorative is because they offer the possibility of relaxing directed attention, giving an opportunity for reflection and thus, 'many of the fascinations afforded by the natural setting qualify as "soft" fascinations: clouds, sunsets, snow patterns' can be enjoyed without effort.[122] In relation to viewing the night sky, this opens up another line of inquiry, namely whether there may be qualitative differences, a different 'felt' expe-

119 Hartig *et al.*, 'Restorative Effects'.
120 R. Kaplan and S. Kaplan, *Experience of Nature*.
121 Ulrich, 'Aesthetic and Affective Response'.
122 Kaplan, 'The Restorative Benefits of Nature', p. 174.

rience, in encountering the night sky through a telescope as compared to naked-eye astronomy.

Unsurprisingly, research involving rural inhabitants reports a stronger sense of connectedness to nature than studies of those living in urban areas.[123] This connectedness can also encourage social interaction and enhance community cohesion.[124] Some studies also suggest the more beautiful a natural scene is perceived to be the more likelihood there will be social interaction amongst those visiting it or living there.[125] Other studies indicate that if humans feel more connected to nature they will feel a greater responsibility to protect it.[126] In an exploration of the mod-

123 R. S. Ulrich, R. F. Simons, B. D. Losito, E. Fiorito, M. Miles, and M. Zelson, 'Stress Recovery During Exposure to Natural and Urban Environments', *Journal of Environmental Psychology* 11.3 (1991): pp. 201–30; Seymour, *Nature and Psychological Wellbeing.*

124 C. L. E Rohde and A. D. Kendle, *Human Well-being, Natural Landscapes and Wildlife in Urban Areas: A Review. English Nature Science Report No. 22* (Peterborough: English Nature, 1994), p. 151; T. A. More, 'The Parks are Being Loved to Death. And Other Frauds and Deceits in Recreation Management', *Journal of Leisure Research* 34.1 (2002): pp. 52–78.

125 Jia Wei Zhang, Paul K. Piff, Ravi Iyerb, Spassena Koleva, Dacher Keltner, 'An Occasion for Unselfing: Beautiful Nature Leads to Prosociality', *Journal of Environmental Psychology* 37 (2014): pp. 61–72.

126 F. Stephan Mayer and Cynthia McPherson Frantz, 'The Connectedness to Nature Scale: A Measure of Individuals' Feeling in Community with Nature', *Journal of Environmental Psychology* 24.4 (2004): pp. 503–15; E. K. Nisbet, J. M. Zelenski and S. A. Murphy, 'The Nature Relatedness Scale: Linking Individuals' Connection with Nature to Environmental Concern and Behavior', *Environment and Behavior* 41 (2009): pp. 715–40; P. W. Schultz, 'Inclusion with Nature: The Psychology of Human-Nature Relations', in *Psychology of Sustainable Development*, P. W. Schmuck and W. P. Schultz, eds., (Norwell, MA: Kluwer Academic, 2002), pp. 62–78.

ern environmentalism movement, Eckersley suggests an enhanced awareness of this connectedness can lead to individuals becoming more politically active.[127] Conversely, being in nature can be so familiar for some that it is taken for granted and lead to less motivation to getting involved in preserving it, an area S. Kaplan explores.[128]

In addition to the possible beneficial effects of nature experiences, authors such as John Davis, Jorge N. Ferrer and Mathews have all independently argued that encounters with nature can lead to positive transformative experiences, which could be of a spiritual or religious kind.[129] In contrast, Sebastiano Santostefano suggests a transformative experience would not automatically be positive, particularly if a person's early experiences of being in nature have been problematic or not well integrated into their sense of self.[130] Mathews and K. Williams and D. Harvey separately describe the potential transpersonal dimensions to nature experiences, such as feelings of identifying with the environment, of oneness and unity, changes in time perception and alterations to personality as a re-

127 Eckersley, *Environmentalism*.

128 Stephen Kaplan, 'Human Nature and Environmentally Responsible Behavior', *Journal of Social Issues* 56.3 (2000): pp. 491–508.

129 John Davis, 'The Transpersonal Dimensions of Ecopsychology: Nature, Non-Duality and Spiritual Practice', *The Humanistic Psychologist* 26.1–3 (1998): pp. 60–100, p. 69; Jorge N. Ferrer, *Revisioning Transpersonal Theory: A Participatory Approach to Human Spirituality* (New York: SUNY Press, 2002), p. 123; Mathews, *Ecological Self*, pp. 149–51.

130 Sebastiano Santostefano, 'The Sense of Self Inside and Environments Outside: How the Two Grow Together and Become One in Healthy Psychological Development', *Psychoanalytic Dialogues* 18.4 (2008): pp. 513–35, p. 513.

sult of peak and transformative experiences.[131] Research conducted by Rebecca Fox suggests such nature experiences are associated with, 'moments of transcendence and spiritual enchantment'.[132] In a survey of a representative sample of a thousand people in the San Francisco Bay area regarding peak experiences, Robert Wuthnow found 82% of respondents had, 'experienced the beauty of nature in a deeply moving way', with 49% believing this had enduring influence.[133] The value of empirical research into transformative experiences is however challenged by Herbert W. Schroeder who comments these experiences may best be revealed by qualitative accounts which may give a more, 'balanced relationship between the rational and the intuitive sides of the psyche', rather than attempts at quantitative measurements.[134]

Many personal narratives also describe transformative experiences in nature, such as those by Mark Coleman, who describes practical exercises to reconnect with nature, Steven Harper who advocates going into wilderness areas, and Sara Harris who experienced transformation

131 Mathews, *Ecological Self*, p. 162; Williams and Harvey, *Transcendent*, pp. 249–60.

132 R. Fox, 'Enhancing Spiritual Experience in Adventure Programs', in *Adventure Programming*, J. Miles and S. Priest, eds., (State College, PA: Venture Publishing, 1999), p. 455.

133 Robert Wuthnow, 'Peak Experiences: Some Empirical Tests', *Journal of Humanistic Psychology* 18.3 (1978): pp. 59–75.

134 Herbert W. Scroeder, 'The Spiritual Aspect of Nature: A Perspective from Depth Psychology', http://www.nrs.fs.fed.us/pubs/gtr/gtr_ne160/gtr_ne160_025.pdf [Accessed 14 Dec 2013].

on a wilderness vision quest which subsequently informed her therapeutic practice.[135]

In conclusion, whilst many of the writers focus on the benefits to be gained from cultivating a connection to nature, they generally understand nature to be 'earth-bound' and give little consideration to fostering a relationship with the sky. Most of the research in ecopsychology which explores the idea that human beings may have a synergetic, or mutually beneficial, relationship with nature, is either theoretical or descriptive in character rather than based on direct measurement and observation of human experience. The fields of health and environmental psychology are more helpful in pointing the way as to how to go about measuring the restorative and transformative effects which may result from contact with nature. Literature from cultural astronomy, the Dark Sky movement, astronomical tourism and personal testimony from individuals give further insights into how the night sky may affect the lives of human beings. I have also highlighted that although most people recognise the benefits of spending time in 'green/grounded' nature, for some people the promotion of dark skies may not be viewed as altogether a good thing. It may lead to questions about the cost of replacing street lights, trigger painful memories, perpetuate fear of the dark, lead

135 Mark Coleman, *Awake in the Wild: Mindfulness in Nature as Path to Self-Discovery* (San Francisco: Inner Ocean Publishing, 2006), p.33 and p. 221; Steven Harper, 'The Way of Wilderness', in Roszak *et al.*, *Ecopsychology*, p. 185; Sara Harris, 'Beyond the "Big Lie": How One Therapist Began to Wake Up', in *Ecotherapy: Healing with Nature in Mind*, Linda Buzzell and Craig Chalquist, eds., (San Francisco: Sierra Club Books, 2009), pp. 87–89.

to concerns about increased crime, loss of privacy, etc. In other words, although 'no man (*or woman*) is an island', as most of us live in a geographical community, there may sometimes be competing priorities at play. Exploring the role of the night sky on an island where the benefits of contact with nature is taken for granted and there is a strong drive to preserve the unpolluted dark sky therefore offers a unique opportunity.

CHAPTER 3:

SHARPENING THE FOCUS: SETTING UP MY RESEARCH

RESEARCH STRATEGY

BEFORE I SET off to interview the residents of Sark, I put a great deal of thought into the best way of going about my wider research project, as is expected of any good researcher. I needed a strategy. How exactly, I wondered, could I find out how the people of Sark feel about the sky? There are many different ways of conducting research, and they all have their individual strengths and weaknesses. While I considered my options, it was important to me to play to my own strengths as a listener. I was clear that my interviews would take place face-to-face. I was interested in visual clues, body language and gestures.

In the end, I adopted the method of intuitive inquiry developed by two American academics, Rosemarie Anderson and William Braud.[1] It is not one of the better known research methods, and indeed does not feature on many postgraduate syllabuses, but I was attracted to it because it seems to allow a greater acknowledgment of the impact of qualitative research on both the researcher and the participants. Moreover, my aim was not only to reach a deeper understanding of the participants' felt experience of encounters with the sky, but also along the way to reflect on my own experience of Sark's dark sky. These days it is increasingly recognised that the researcher is actually a part of the research process.

Intuitive inquiry is informed by values and practices from the field of transpersonal psychology, that is, the school of psychology that is interested in experiences. As Roger Walsh and Frances Vaughan said, 'the sense of identity or self extends beyond (trans) the individual or personal to encompass wider aspects of humankind, life, psyche or cosmos'.[2] Transpersonal approaches allow for innovative methods of gathering information about human experiences which recognise alternative methods of awareness and intuition throughout the actual research process, rather than dismissing them.[3] Dreams, synchronicities, chance encounters and what William Braud calls 'gut feel-

1 William Braud and Rosemarie Anderson, *Transpersonal Research Methods for the Social Sciences* (Thousand Oaks, CA: Sage Publications, 1998), p.69.

2 Roger Walsh and Frances Vaughan, 'On Transpersonal Definitions', *Journal of Transpersonal Psychology* 25.2 (1993): pp. 125–82.

3 Braud and Anderson, *Transpersonal Research Methods*, p. ix.

ings' are therefore all seen as valuable and necessary if we really want to know how people feel. By using intuitive inquiry I could employ personal and clinical skills with which I was already familiar, such as reflective listening, working with intention and various forms of intuition.[4] I have been practicing as a transpersonal psychotherapist for eighteen years and am accustomed to working in this way. Furthermore, intuitive inquiry takes account of the possibility of research providing opportunities for the transformation of the researcher – me – and participants and readers of the research.[5] After all, people are changed by doing research, engaging with it and reading about it. In other words, there may be long-lasting benefits for everyone involved in the process – an exciting prospect!

This might sound very woolly but, doing intuitive research is actually quite challenging, as academics have recognised. Anderson pointed out that if the research is to be successful, the researcher must be, 'rigorously aware of one's internal processes or perspective', and at each stage of the research they must evaluate what they have learned and what they feel.[6] I found this expectation quite

4 William Braud, 'Towards a More Satisfying and Effective Form of Research', http://contemporarypsychotherapy.org/vol-2-no-1/towards-a-more-satisfying-and-effective-form-of-research/ [Accessed 16 Feb 2014].

5 Rosemarie Anderson and William Braud, *Transforming Self and Others through Research: Transpersonal Research Methods and Skills for the Human Sciences and Humanities* (New York: State University of New York Press, 2011), p. xv.

6 Rosemarie Anderson, 'Intuitive Inquiry: An Epistemology of the Heart for Scientific Inquiry', *The Humanistic Psychologist* 32.4 (2004): pp. 307–41, p. 307.

challenging; although I felt within my comfort zone when I was conducting the interviews, some of the issues that participants raised caused me to question my involvement in promoting dark skies in my hometown. I asked myself whether I wanted to become more involved, or whether I had competing priorities of my own, not the least of which was writing up my research! Conducting the interviews on the island (I could have done them by email or Skype) and allowing myself to have a personal experience of Sark's night sky proved valuable contributions to my research, as they facilitated my own reflexive process. I was able to respond to the same sky that I was asking people about, sharing something of their experience. I kept a reflexive journal during the course of my research. I followed the best academic advice from Michele Ortlipp who said that 'keeping self-reflective journals is a strategy that can facilitate reflexivity'.[7] Anderson and Braud added that keeping a reflexive journal can act as a check and balance for qualitative findings.[8] What they mean is that data collection is not a totally objective process, but reflects the researcher's feelings, expectations and preconceptions – in this case, mine. I set about compiling a record of my own reflections, dreams and notes about significant events. On several occasions, I even wrote my journal entries outdoors on moonlit nights on Sark!

7 Michelle Ortlipp, 'Keeping and Using Reflective Journals in the Qualitative Research Process', *The Qualitative Report* 13.4 (2008): pp. 695–705, p. 695.

8 Braud and Anderson, *Transpersonal Research Methods*, p. 214.

For the project I used a method called 'snowball sampling' to identify and select possible participants living on Sark for interview.[9] The researcher starts with one contact and connections then snowball, as the initial contact makes referrals to other potential participants, who go on to make referrals to additional participants, and so on.[10] My first contact was my friend Alex who agreed to recruit participants through his extensive social network. These were people who he believed, in consultation with me, would be interested in participating and whose participation would be likely to contribute to the research project's objectives. I had also met several of them on my previous trips to Sark, so they were not total strangers. I needed a selection of different kinds of people: some who had lived all their lives on Sark (*Sarkese*), some who travelled regularly to other places and others who had come to live on the island more recently. One participant, who is a member of the Sark Astronomical Society (SAstroS), also suggested a group meeting with some Society members, a suggestion I eagerly took up. I hoped that this mixture of people would allow me to obtain a sufficient variety of different perspectives.

As I have a small number of friends on the island and have visited several times before, one could consider me to have some degree of 'insider' perspective which may assist the researcher to better appreciate what we know in

9 Rowland Atkinson and John Flint, 'Snowball Sampling', http://srmo.sagepub.com/view/the-sage-encyclopedia-of-social-science-research-methods/n931.xml [Accessed 16 Feb 2014].

10 Diane C. Blankenship, *Applied Research and Evaluation Methods in Recreation* (Champaign, IL: Human Kinetics, 2010), p. 88.

research parlance as 'the individual actors'.[11] As I live at some distance from Sark (Edinburgh is 452 miles away), my chosen sampling method enabled me to enlist participants whom I would not easily be able to contact face-to-face myself. Beginning with Alex, a network was created. Of course, a disadvantage of this method, as previous researchers have pointed out, is that there is no way of knowing whether my relatively small sample is representative of the larger population of Sark or elsewhere.[12] Also, I was one step removed from the recruitment process and the choice of who was recruited was largely dependent on the judgement of Alex and others, and consequently was subject to any bias in their judgement. However, the advantage of small samples is that we can focus on individual stories rather than aiming for some mass opinion which might not exist anyway. The reality was that every single person Alex approached was interested in taking part, and this demonstrates for me the high level of interest amongst Sark residents regarding their dark sky.

11 Kenneth L. Pike, 'Etic and Emic Standpoints for the Description of Behavior', in *The Insider/Outsider Problem in the Study of Religion. A Reader*, Russell T. McCutcheon, ed., (London: Continuum, 1999), p. 32.

12 Patrick Biernacki and Dan Waldorf, 'Snowball Sampling Problems and Techniques of Chain Referral Sampling', *Sociological Methods and Research* 10.2 (1981): pp. 141–63, p. 160.

GATHERING THE DATA

During the two months prior to my visit to Sark, I refined the key questions to be covered in my interviews. First of all, I conducted trial interviews with several people, including Alex himself, and reworked certain questions as a result. The original list of questions proved to be very long and would have required allocating at least an hour-and-a-half for the interviews. That is too long in most circumstances. My interviews were semi-structured: I had an initial format but with space to vary it if required. I asked some specific questions, as I wanted to gather data on specific areas, for example, time living on Sark and time spent away from Sark. But I was also keen that I allocate enough time in the interviews for people to elaborate on any areas that they felt were important. I was particularly interested in five main issues: the possible innate human desire to see the night sky; the commercialisation of this desire through astronomical tourism; fear of the dark; the 'nature' of nature; and nature and wellbeing. In total, I identified ten potential participants, all of whom agreed to be interviewed. However in the end two were unable to participate due to other commitments, so I had eight key interviewees.

In March 2014, I visited Sark for five days with my partner, staying with Alex in a house just off the Avenue, one of the island's main streets. I interviewed my eight participants in locations of their own choosing, for anywhere between thirty five to sixty minutes, less than some of my initial trial interviews, but long enough for people to tell me what they wanted. My information as supplemented by a small, informal, focus group with three members of SAstroS at the Island Hall. I had not originally intended to hold such a group but many researchers recommend them on the grounds that they give what they call 'concentrated and detailed information on an area of group life'.[13] This was precisely what I needed. They are actually an extremely important feature of modern political management, so very much part of modern life. However, probably unlike the average political consultant, I was aware that I was not a detached observer. Still, it was a valuable opportunity for people to tell their own stories.

Having emailed members of SAstroS in order to alert them to the focus group, Annie Daschinger, the society's chair, also emailed the interview questions to three further members who had been interested but unable to attend; they subsequently sent me very valuable and detailed comments. My entire group therefore comprised fourteen people, of whom nine were women and five were men. In my original MA dissertation all material from the participants was cited anonymously. This was an ethical requirement. However, sometimes people like their views to be public so I contacted everyone again and asked if they would be willing to waive anonymity in order that I might give more 'local colour' and better contextualise their comments in the book. Ten people consented, one declined, one person unfortunately had since died, and two either did not

13 Michael Bloor, Jane Frankland, Michelle Thomas and Kate Robson, *Focus Groups in Social Research* (London: SAGE, 2001), p. 6.

respond or their email address was defunct. Those whom I could not contact have been given pseudonyms.

Whilst some participants were retired (but busy!), others worked in a variety of one or more occupations, including teaching, finance, construction, gardening, art and photography. Alex White, who had moved to Sark from Edinburgh, regularly travels to London and works in financial services. Jeremy La Trobe-Bateman is the current Seneschal of Sark and one of the seven members of SAstroS who took part. He grew up in England but his grandparents lived on Sark and he always visited for holidays. Rosie Byrne is a conseiller (a member of the Sark parliament) and Sarkee, from one of Sark's oldest families, who has been painting and drawing the island and its wildlife for over thirty years. Mandy Grey is the Class 3 teacher at Sark School and ran an astronomy group at the school for a time. Lydia Bourne is a photographer, a visitor officer for Sark Tourism and is also responsible for polishing the optic at Point Robert Lighthouse in the north of the island. Annie Daschinger is an artist, white witch and Chair of SAstroS. Reg Guille is a long-term resident of Sark and President of the Chief Pleas. Puffin Taylour organises Sark's annual sheep racing contest and has regularly watched stars from her cabin in Greenland. Gerry Loughlin is a builder and originally from Ireland. Prior to coming to Sark he built his own house in rural Ireland in an area where the sky was very dark. Roz Rolls, a musician and gardener, is from New Zealand. Some participants have lived all their lives on Sark, others have come from both urban and rural areas elsewhere – either because of work, in search of a different way of life or to be with a partner. The time participants have spent living on Sark ranges from two years to more than forty years. Several people were already familiar to me from episodes of, 'An Island Parish': in one episode Puffin Taylour encouraged people to participate in Sark's annual scarecrow competition; in another, Lydia Bourne, along with Sue Daly, persuaded the island's emergency services to strip off and be photographed for a fundraising calendar; in a third Jeremy La Trobe-Bateman led a volunteer work party to repair the path down to Grande Greve beach.

I also read through the letters of support included in the island's original application to the IDA for Dark Sky status as they helped give me a broader picture of the residents who were committed for various reasons to preserving the night sky on Sark.[14]

I ended up with a huge amount of material from my interviews, the focus group, my reflexive journal and various emails. Gradually, I began the process of transcribing the recorded data manually onto paper, making notes and occasional drawings in the margins to capture my initial 'take' on the data. Suns, lightbulbs and stars put in regular appearances! Sometimes, I also noticed a particular image or feeling coming up for me as I transcribed and I noted it down. While writing up the data relating to 'fear of the dark' for example, I personally experienced a physical sense of unease in my stomach; that night I dreamt of sharks circling a small island. One night I woke with a sense that there were fears being expressed implicitly by

14 Steve Owens, 'Sark Dark Sky Community. A Dark Sky Island'.

some participants of outside negative influences on their way of life. I proceeded with caution however; as Anderson wrote, 'like observational data, intuitions oblige corroborative evidence since they are subject to error and bias'.[15] The last thing I wanted was for my intuition to lead me down the wrong path. However, I was pleased that my intuition appears to be corroborated by comments from Alex, '(outsiders) [...] who see a different sky to us' and Rosie, who saw some outside individuals as 'elements [...] that are a constant threat'. Sark and shark are very similar words! Indeed, as I described earlier there are outside pressures having an impact on Sark's way of life.

Once I was back in Edinburgh, I set out to understand my data. I used thematic content analysis, a method which can be useful in identifying, organising and describing themes within qualitative data, as it aims to balance the more subjective aspects of transpersonal research.[16] I read through my text highlighting all descriptions relevant to my research topic, and then I grouped them together as distinct themes, or 'units of meaning', which I then copied onto individual pieces of paper. I spent some time reflecting in my journal on the large number - forty three in all - of themes I identified. I wrote 'as I contemplate what lies around me I feel as I have done at times when gazing at

15 Anderson, *Interpreting*, p. 2.

16 Rosemarie Anderson, 'Thematic Content Analysis (TCA): Descriptive Presentation of Qualitative Data', http://www.wellknowingconsulting.org/publications/pdfs/ThematicContentAnalysis.pdf [Accessed 14 Feb 2014]; R. E. Boyatzis, *Transforming Qualitative Information: Thematic Analysis and Code Development* (London: SAGE Publications, 1988), p. vii.

a starry sky – what first seems a mass of individual stars until gradually patterns and constellations materialise'.[17] I allowed myself a period of 'incubation' between work sessions, so I refrained from looking at the piles arranged on the floor in my conservatory for several days. Then I put similar units together and re-labelled and sub-divided others. Eventually I came up with a final list of seven themes, which comprised the themes I had originally identified, but also incorporated some additional ones which I had not anticipated. For example I labelled one unexpected them, 'Observing the night sky with others as a means of building and maintaining family/community connection'.

A particular feature of intuitive inquiry is the way in which researcher's initial understanding of the topic is transformed by incorporating the perspectives of others, having what Anderson called 'intuitive breakthroughs'.[18] Perhaps the most impressive thing I came to realise about myself was that I no longer feared going out alone on a dark night on Sark, as I trusted my night vision would 'see me through', having what Anderson called Limitations and Problems

My original intention had been to conduct twelve interviews, but I quickly came to realise that this would involve transcribing too large an amount of material in the allocated timeframe. Most research is conducted within a limited timeframe, so I decided to reduce the number to eight. I felt that this, in addition to material from the fo-

17 Direct quote from author's personal journal.
18 Anderson, *Epistemology*, p. 321.

cus group, would be sufficient to gather enough information. Judging by the volume, breadth and depth of relevant material subsequently gathered, this feels to have been the right decision.

I had originally intended to travel to Sark twice for the interviews, but in fact I only visited once because I managed to conduct more interviews during my March visit than I had anticipated. However, this did mean that I had less time to immerse myself in the Sark 'experience'. I also prepared by reading Mervyn Peake's fictional account of a visitor to Sark, *Mr. Pye*, and Rosie's, *Sark Sketchbook: Journal of a Local Artist*.[19] Yet, throughout the research it was always my intention to return to Sark after I had written up my findings and give some kind of public presentation. That opportunity finally arrived in September 2015. My friend Alex took the lead in advertising and organising the event, and Sue Daly, a local photographer, generously contributed a photograph for the poster for my talk in the Island Hall to around fifty people. John Hunt, the caretaker at the Hall who had help me set up my laptop, was also one of the island's firemen, multitasking as is common on Sark! The audience included several of my interviewees, other islanders and a few visitors, and some of the topics raised at the event are included in the following discussion. The day after the talk, an islander who had not been able to attend but had heard about my research, struck up a conversation with me on the Avenue and asked me a question I had

19 Peake, *Mr. Pye*; Rosanne Guille, *Sark Sketchbook: Journal of a Local Artist* (Sark: Cat Rock Publications, 2004).

never considered before: Had I interviewed myself using the same questions as my interviewees? In fact, I had not! However, following this suggestion, I interviewed myself using these same questions and recorded the responses in my journal.

CHAPTER 4:

BIFOCAL VISION: MY RESULTS AND DISCUSSION

WHEN I WAS training in psychosynthesis, a transpersonal therapy, I learnt about the notion of what practitioners call 'bifocal vision'. This is a way of attending to the detail of what someone is communicating to you even while at the same trying to get a sense of the deeper layers of meaning, expanding the context of what is being communicated. The final stage of my research process involved integrating the findings from the interviews, focus group and email responses with material from my original review of the literature. As I listened to the different thoughts and feelings participants shared with me about Sark's night sky I attempted to understand as clearly as possible not only what was being said about themselves and their life on Sark, but also what messages there might be for those of us living in a less close relationship with the night sky. Eventually, seven main themes arose and, as I describe below, a number of new and unexpected themes emerged relating to memories of childhood sky experiences, the experience of watching the night sky with others and fearlessness of the dark. The rest of the chapter is organised around these themes, as I found them expressed by my participants in my research.

MEANING AND SIGNIFICANCE: THE HUMAN DESIRE TO CONNECT WITH THE NIGHT SKY

Earlier I described William E. Kelly's work on the concept of *noctcaelador* or intense attachment to the night sky.[1] I found my interviewees all expressed such sentiments. Jeremy was on who illustrated this: 'I have a great love of the night sky', he said. Rosie who used to go on night fishing expeditions around Sark as a child, told me, 'I absolutely love it [...] can't imagine being without it'.[2] Some participants also demonstrated a broad spectrum of astronomical knowledge and expressed a range of opinions regarding the value or need for this type of knowledge. 'Paula', a botanist and former night carriage driver on Sark, who first came to live there forty years ago, thought sky knowledge could enhance the experience,

1 Kelly, 'Development', pp. 100–2.
2 Direct quotes from Jeremy and Rosie interviews.

> it's a bit like the dawn chorus (*the night sky*), you can listen like you could to an orchestra, it's just a load of sound which is very pleasing but the more you understand that's a blackbird, that's a robin, that's a thrush, the more you start getting it, appreciating the instruments. Sometimes it's nice just to forget all that and use your ears and enjoy and it's the same as the night sky. Obviously if you can name constellations and think about things that adds to it.[3]

'Miranda', by contrast, who had been on Sark for about fifteen years, had more limited astronomical knowledge, but still connected powerfully with the sky: 'I'm constantly aware of the night sky and what's happening [...] the only thing I recognise [...] there are three stars in a row [...] I don't know their names and I'm not interested'.[4] She added that although she would on occasion comment on the sky with others, she was wary of sounding, 'too excited, enthusiastic' as others might say, 'it's just the sky, it's always dark – what are you getting excited about?'[5] She preferred more solitary stargazing. Both Paula and Miranda's quotes support Kelly's assertion that it is not necessary to have detailed astronomical knowledge to have a close relationship with the sky.[6] Conversely astronomical knowledge may sometimes get in the way of connecting with the sky as Annie found when she asked some, 'fancy astronomer' visit-

3 Direct quote from Paula focus group.
4 Direct quote from Miranda interview.
5 Direct quote from Miranda interview.
6 Kelly, 'Development', pp. 100–2.

ing Sark, 'can you tell me what's up there? He said, "I don't know, the trouble is I'm so used to looking with my telescope I don't look with my eyes at the sky'.[7] Puffin thought that many visitors to Sark had 'forgotten how it (*the night sky*) looked when they were younger and never bother to look now', thus missing 'an important aspect of their life'.

The degree of astronomical knowledge people possessed appeared to bear no relation to how much time was spent outdoors or how long they had spent living in light-polluted areas. In fact Mandy, who had lived in cities large parts of her life and described moving to Sark as, 'a bit of a shock to the system', probably had the most astronomical knowledge.[8] Clearly, light pollution is not a necessary bar to acquiring knowledge about the night sky. For most SAstroS members, such as Jeremy, looking at the night sky was a purposeful act, regularly undertaken either alone or with others: 'if it's a clear night I'll always stick my nose out and see what's doing up there', and Puffin appreciated that 'Annie points out things via text at special times'.[9] But others, such as Gerry, also took the time to do so. He told me that 'when I've gone out for logs or coal at night, I'll always stop for a minute and have a look'. For some, looking at the sky was done in between doing other things: as Mandy remarks, it 'tends to be when you're on your way home'.[10] This

7 Direct quote from Annie focus group.
8 Direct quote from Mandy interview.
9 Direct quote from Jeremy interview and Puffin email.
10 Direct quote from Mandy interview.

is a kind of what Kaplan called 'soft' fascination, enjoying nature without effort.[11]

Sometimes, though, even on Sark the night sky could disappoint. Annie and Paula relating a memory from 2013 when 'We were disappointed about (*comet*) Ison' because 'we waited with baited breath'.[12] Their expectations were not met! More recently Reg missed the Northern Lights, 'I damn well didn't see them!', he complained, adding, 'People on Guernsey and Jersey saw them!'[13]

Furthermore, a number of participants, such as Alex, felt some people took the Sark night sky for granted – 'I think older folk take it for granted, don't appreciate it in the same way', and Rosie advised me, 'it's not something we talk about all the time'.[14] An article on stargazing on Sark Tourism's website suggests this is a common view 'People who live with myriads of bright stars', the site said, 'tend to take them for granted, a mere adjunct to all the other beauties of island life'.[15] On Sark as elsewhere, some people lead busy lives and feel that they don't have time to look up and watch the sky. As Mandy says, 'I don't go out often just to have a look, there's always things you need to do'.[16] Stephen Kaplan proposes that those who regularly spend time in 'green' nature may take it for granted and not feel motivat-

11 S. Kaplan, 'The Restorative Benefits of Nature', pp. 172–74.
12 Direct quotes from Annie and Paula focus group.
13 Direct quote from Reg focus group.
14 Direct quotes from Alex and Rosie interviews.
15 Sark Tourism, 'Star Gazing', http://www.sark.co.uk/star-gazing-12757/ [Accessed 12 Feb 2016].
16 Direct quote from Mandy interview.

ed to preserve it.[17] Perhaps for some long-term residents of Sark the sky had become 'part of the furniture', and it was assumed that it would always stay the same. These, though, weren't the people who spoke to me.

In contrast, for some people the night sky held special meaning and significance. Roz who has been on Sark for almost ten years and previously lived in a number of different countries reflected on the Pleiades, one of the most distinctive sights in the night sky:

> the seven sisters, in New Zealand they call [them] *Matariki* and for me that's quite a special constellation. From here it looks like the shape of New Zealand......When I see them here it's really special, I say, 'that's my New Zealand stars' [...] the dark sky makes me feel like praying or something.[18]

The Maori New Year is marked by the rise of the *Matariki* and is traditionally the time of year when new crops are planted. For Lydia, a mother with young children, watching the night sky was a way of staying connected to her own mother,

> my mum [...] passed away five years ago [...] a certain relief when you look at the sky, maybe trying to connect with somebody you've lost. I talk to the sky quite a lot [...] it can make you feel very small, the vastness of it.[19]

17 S. Kaplan, *Human nature*, pp. 505–6.
18 Direct quote from Roz interview.
19 Direct quote from Lydia interview.

Both Lydia and Roz saw the dark sky offering a way of maintaining a connection to a loved one or a faraway birthplace, a form of meditation or prayer. Most participants did not consider themselves religious, but the language used by some to report their experiences closely mirrors that found in some religious sentiments. For example, Rosie commented that, 'you kind of feel small, you feel your place in the universe'. Jeremy told me that 'the night sky definitely makes you more aware you're part of a great creation'.[20] In my own journal I noted, 'it's like when I lit candles in church in the darkness for people I've lost [,] only they don't need candles on Sark [,] they have their stars'.[21]

Turning now to sky stories, although Anthony Aveni comments that humans have always employed imagination to create imagery and stories about the sky, other than one participant's reference to the book *Sark Folklore*, no participants recalled ever reading about or hearing any stories relating to the sky specific to Sark.[22] Even Rosie, whose ancestors lived on Sark, reported that he had 'never come across any on Sark, we're very non-spiritual. Everybody's so busy all the time there's probably never been time to think up stories about the stars [...] I wasn't passed down stories about the night sky'.[23]

Several longer-term and also newer residents mentioned hearing tales of witches and Jeremy commented

20 Direct quotes from Rosie and Jeremy.

21 Direct quote from author's personal journal.

22 Aveni, *Conversing*, p. xiii; Remphry, *Sark Folklore*.

23 Direct quote from Rosie interview.

that, 'there's a very strong folk history on the island but it's more related to witches and creatures'. Perhaps he had in mind creatures such as the Witch's Hound, the *Tcicho* reputed to lurk near La Coupee and who reputedly chases those about to die. Martin Remphry notes that a belief in witchcraft was prevalent on Sark until the twentieth century, and Alex mused, 'there's a lot of stories about witches so maybe there's some star story in there'.[24] Amongst the majority of participants there was an assumption that Sark sky stories must exist and older people or members of SAstroS would know them. However, I heard none.

Not long after the interviews Sark Theatre Group mounted a production of Shakespeare's *Macbeth*, a story involving witches! Jeremy was aware of the archaeological excavations which had taken place on the island and speculated, 'the ancients who lived on Sark – I've always imagined those people were very connected to the sky [...] there were lots of dolmens'.[25] Echoing Jung's remarks regarding UFOs being a 'living myth', Annie, commented sanguinely, 'I think the modern myths are the UFOs now'.[26] I reluctantly came to the conclusion that if there ever were any Sark stories then they are now lost, and this reflects the loss of sky stories in the community.

My disappointment regarding the absence of Sark sky stories disappeared after re-reading the interview tran-

24 Remphry, *Sark Folklore*, p. 13; direct quote from Alex interview.

25 Direct quote from Jeremy interview.

26 C. G. Jung, *Flying Saucers: A Modern Myth of Things Seen in the Sky* (Princeton, NJ: Princeton University Press, 1979), p. 14; direct quote from Annie focus group.

scriptions over several days when I finally had an 'intuitive breakthrough'. Just as previously 'hidden' stars and planets are becoming gradually known to astronomers, some themes I had anticipated might appear, but had not initially found, emerged under different guises. I wrote, 'all these interviews are Sark sky stories! Everyone has their own individual tale to tell' – this also includes me, I realised I too had Sark sky stories.[27] It was a mistake to imagine that stories have to be handed down from the distant past: we are still creating them everyday.

During my visit in September 2015 I attended the island's first ever Son et Lumière production, *Fantasie*. This outdoor show depicted Sark's history from its geological formation millions of years ago to the present day. It was part of the celebrations marking 450 years since the signing of the Charter that declared Sark to be a fief, and it was a sell-out. Likely many cultural enterprises on Sark, it involved the voluntary efforts of more than fifty locals, all of whom were willing to attend rehearsals, sell tickets, and so on. The magnificent setting of the seventeenth century La Seigneurie (the official home of the present Seigneur), its formal gardens and the clear skies overhead made for a stunning spectacle. Seeing the constellation of the Plough vividly etched against the sky amongst a forest of stars too numerous to identify is an image I will always remember. And yet the previous night, I had imagined that nothing could better than walking along the quiet road near my friends' house and seeing the Milky Way arching from ho-

27 Direct quote from author's personal journal.

rizon to horizon. One blot on the horizon (literally!) for me was the finale of the performance, when the tower at the top of La Seigneurie was illuminated and some fireworks were let off. This resulted in all the stars disappearing from view, much as they do when there is a lot of light pollution. I wished the natural bright lights above had been left; this would have been stunning enough. The day after the performance, people were recounting to others who had not been able to get a ticket how beautiful the clear night sky had been – in other words, telling a Sark sky story.

Another very different outdoor event I was invited to that weekend was a birthday celebration of some people my friends knew. The party took place on a beautiful hill gently sloping down towards the sea in the quieter south end of the island. A small toilet tipi had been erected with an open entrance giving beautiful views of the sea and twilight sky. As the sun began to set some guests who weren't from Sark commented on how it seemed to slide across the sky rather than dip towards the horizon. Those with more local knowledge said this was what it always did from this side of the island and to expect it to suddenly seem to dive down and disappear from sight. As we watched it did just that: another Sark sky story.

Many participants had shared experiences of particular sky events they had witnessed on Sark with their children, and thus it seemed these stories were being handed down through the generations. Reg, who was born and brought up on Sark and whose family have always lived there, relates his own childhood story and a recent one. He told me,

> I saw the Northern Lights on Sark as a child [...] out playing and low and behold the sky to the north just lit up, we had no idea what it was, absolutely wonderful [...] in August last year, the shooting stars, they were brilliant. My family, the grandchildren, were all here and we were all out in the garden [...] everybody was looking in different directions and saying did you see that one?[28]

I was reminded of Paul Bogard's observation that dark night skies are a 'continuing source of stories that help us understand and live our lives'.[29]

COMMUNITY: OBSERVING THE SKY WITH OTHERS AS A MEANS OF BUILDING AND MAINTAINING FAMILY/COMMUNITY CONNECTION

The enjoyment and high value the majority of participants attached to watching the night sky on Sark with family, friends and others was an unexpected finding. Roz comments,

> you notice it every night when you're a smoker, especially if you're with other people, you're looking at it together, pointing things out [...] when more than one of you see it (*a*

28 Direct quote from Reg focus group.
29 Bogard, *Let There Be Night*, p. 86.

> *shooting star*) at the same time that's special [...] you discuss the sky, 'is that Mars, is that Venus'? [30]

(In fact, I saw more shooting stars during my first trip to Sark than I had seen in my lifetime). Gerry, who had originally come to Sark to work for a few months and had stayed on as he enjoyed living there so much, also speaks about this social aspect amongst a temporary community of workers,

> myself and a few friends, a lot of men living together because we were renovating the hotel here, often after the pub we'd pull the mattresses out and then lie on the field and watch the stars and the satellites. We counted thirty satellites one night.[31]

People who sometimes mistake planes from stars are not so far off the mark, if star watchers now include artificial satellites. I am reminded of Clive Ruggles' and Nicholas Saunders' remark that what is observed in the sky always has both astronomical and cultural meaning.[32] As mentioned above, memories and sky stories regarding specific sky events were created and passed on, particularly when there were unexpected or unusual occurrences. Jeremy described the 'amazing sight' of seeing a comet, 'coming out of the north... a flaming object and the tail was flaming, smoking, all that debris, some of it coming our way',

30 Direct quote from Roz interview.
31 Direct quote from Gerry interview.
32 Ruggles and Saunders, *Astronomies and Cultures*, p. 1.

and Reg remembered a recent particularly fortuitous night when, 'you could actually see the Orion nebula with the naked eye and I've never seen that before'.[33] Such observations become community events. Several people recalled observing the same things from Sark: Annie, Mandy, Lydia and Reg all vividly remembered the meteor showers in August 2013; Annie, Rosie and Reg had all watched the total solar eclipse in 1999 and heard the tooting of horns from dozens of boats when the sun reappeared and Alex, Jeremy and Gerry spoke of the regular passing overhead of the International Space Station. The common at L' Eperquerie on the northern tip of the island was a favourite place for many to watch such events and Rosie had gone there to do some watercolours during the 1999 eclipse. For Mandy sharing the experience of watching the Perseids meteor shower in the field outside the Island Hall was especially significant,

> last summer lying on the field outside the Island Hall with everybody looking up during the meteor shower [...] people had sleeping bags, thermoses and hot chocolate [...] it was brilliant, you had all the oohs and aahs [...] loved I wasn't doing it on my own, I was with other people [...] everybody was getting so much enjoyment out of it.[34]

Perhaps this communal skywatching is particularly important in maintaining bonds in a small island community where people depend on each other. Also, as there is much

33 Direct quotes from Jeremy and Reg interviews.

34 Direct quote from Mandy interview.

to do on the island and therefore many people have several jobs, the number of people encountered regularly can make up a high proportion of the island's population and may lead to increased opportunity to share experiences. As a number of scholars have observed, spending time in nature can encourage social interaction and enhance community cohesion.[35] And the sky is part of nature.

WELLBEING: EXPERIENCING POSITIVE FEELINGS THROUGH OBSERVING THE NIGHT SKY

As mentioned earlier, apart from Rachel Kaplan's work on the psychological benefits of looking out of a window, research regarding any relationship between experiences in nature and positive outcomes for health and wellbeing does not usually include references to the sky.[36] Various dark sky organisations note the benefits to human wellbeing of a dark night sky but give no evidence to support these claims. I can now provide oral testimony to support these claims, and participants in my study offered many examples of enhanced wellbeing. Comments such as 'relaxing', 'comforting' and 'gives perspective' were frequently given with explanatory descriptions. Out at night Puffin enjoyed, 'Tranquility, the mood to be still and let thoughts come in' and Alex felt that moving to Sark had a positive effect on his

35 Rohde and Kendle, *Human Well-being*, p. 151; More, *Parks*, pp. 52–78.
36 Rachel Kaplan, *Nature of the View*, pp. 507–42.

wellbeing and believed 'the sky must play a part in that'.[37] 'Martin' who confessed to being unaffected, 'emotionally/physically/spiritually' nevertheless felt that being insignificant against the larger canvas of the night sky was comforting: 'awe when you consider what a tiny insignificant dot we are when you consider the vastness of what is out there'.[38] Mandy enjoyed telling her pupils, 'you're made of stardust and if you can get your head around that you're part of something amazing'.[39] Others, like Miranda, deliberately sought out these positive experiences, 'when you've had a bad day, I've gone out into it, gone for a walk'.[40] Mandy remarked, 'this huge mass of stars in the sky, it makes you feel a lot better [...] you look up and you look out [...] it just draws you out, you concentrate on something else...' Paula experienced positive feelings more spontaneously: 'coming back from the pub at night [...] just to stop and look at the stars, feels great'.[41] As Tim Edensor and Shepherd Bliss had claimed, darkness has the potential to promote positive transformation and psychological healing.[42] Again, all that previous research into the restorative effects of spending time in nature is supported.[43] For Alex however who had been dealing with various losses and changes it also

37 Direct quote from Puffin email and Alex interview.
38 Direct quote from Martin email.
39 Direct quote from Mandy interview and Martin email.
40 Direct quote from Miranda interview.
41 Direct quote from Paula focus group.
42 Direct quotes from Lydia and Mandy interviews, Edensor, *Gloomy City*, p. 13; Bliss, 'In Praise of Sweet Darkness', p. 174.
43 Hartig *et al.*, 'Restorative Effects', pp. 3–26; R. Kaplan and S. Kaplan, *Experience of Nature*; Ulrich, 'Aesthetic and Affective Response'.

brought up more poignant feelings, 'The distances, when you're seeing a light and it's now dead, that's quite sad. The thing you are seeing is actually no longer there'.[44] So long does light take to teach us from distant stars that we can never know whether what we are looking at still exists.

A small number of participants spoke of being 'transformed' by watching the sky, but they did not qualify what this meant for them other than it had been profoundly significant as when Alex says, 'it was quite freaky'.[45] For Jeremy it was difficult to find the words to convey his experience, 'hard to put into words or put a figure on that'.[46] There is also evidence that for some the experience was transpersonal or transcendent in some way. Annie told me she was 'enchanted. There's a definite mystical element to going out on your own at midnight'.[47] My own transpersonal experience on Sark happened one night when I had gone outside to have a quick look at the sky before going to bed. Having spent the day interviewing people I felt exhausted and elated with the huge amount of material I had gathered and wondering how I would ever get to sleep. Looking up, despite the mass of stars, I was immediately drawn to the bright orange/red star Betelgeuse, sitting on the shoulder of the giant constellation Orion. A feeling of deep comfort started to spread throughout my body, I felt supported and let myself relax into the warm glow – it was a wonderful

44 Direct quote from Alex interview.
45 Direct quote from Alex interview.
46 Direct quote from Jeremy interview.
47 Direct quote from Annie focus group.

feeling. As Fox had claimed some experiences of nature are associated with transcendence and enchantment'.[48] My findings definitely support previous research by a series of scholars - Davis, Ferrer and Mathews - who all independently maintain encounters with nature can lead to positive transformative experiences.[49]

Transformative experiences for participants in my study appeared to result from seeing particular sky events, such as comets, but their experiences generally did not appear to have spiritual or religious components. However, although most participants described themselves as not religious, for some the night sky itself was associated with spiritual and/or religious beliefs. This is illustrated by Puffin: 'when I look at the Sark sky I am in awe of the size of the universe. How can people not believe in God when we live in such a wonderful, complex world?' and Mandy, 'you can be in awe of God's creation and I don't see how science can get in the way of that'.[50]

NOSTALGIA: CHILDHOOD MEMORIES OF THE NIGHT SKY

Another unexpected finding was how the Sark night sky conjured up fond childhood memories of night skies on

48 Fox, 'Enhancing Spiritual Experience', p. 455.

49 Davis, 'Transpersonal Dimensions', p. 69; Ferrer, *Revisioning Transpersonal Theory*, p. 123; Mathews, *Ecological Self*, pp. 149–51.

50 Direct quote from Puffin email and Mandy interview.

Sark and elsewhere. Many of my participants' comments are reminiscent of those in Paul Bogard's collection of personal testimonies.[51] Gerry commented, 'I love it, reminds me of my childhood, walking in the dark with my grandparents'.[52] Martin, who recently moved to Sark from a city, recalled his childhood night sky as less light-polluted: 'Sark night sky reminded me of how the skies looked like when I was about six or seven years old and growing up in Herefordshire'.[53] Similarly 'Jay', brought up on Sark, remembered the night sky nostalgically: 'aged about twelve, with my girlfriend, looking up and trying to imagine infinity'.[54] For Miranda it brought up memories of the unknown: 'as a child it was all so mysterious [...] the three stars I used to be able to see when I was going to bed'.[55] Perhaps this is reminiscent of Totton's remarks regarding the waning of a relationship with nature for many people as they grow up.[56] However, Joe Slovick suggests appreciating the night sky may be a means of connecting with the past, either the ancestors of one's own culture or another culture.[57] Appreciating the night sky may, perhaps, be a means of connecting with one's past which relates to my discussion of the transmission of sky stories, a point illustrated by Mandy:

51 Bogard, *Let There Be Night*.
52 Direct quote from Gerry interview.
53 Direct quote from Martin email.
54 Direct quote from Jay email.
55 Direct quote from Miranda interview.
56 Totton, *Wild Therapy*, p. 17.
57 Slovick, 'Towards an Appreciation'.

> I remember that child in me that sat and copied constellations out of books and memorised the zodiac constellations [...] sometimes when I look at the sky here I feel like that kid again and I think how did I let this escape? [...] living here it's reignited something I kind of haven't had since I was a child.[58]

Lydia commented that, 'every time I'm out I look up and I encourage the children to do it as well [...] my daughter was born on a full moon and I called her Phoebe, the moon goddess [...] I talk to her about the moon and what her name means...I notice stages of the moon'.[59] A majority of parents on Sark send their children to secondary schools in England and Guernsey at the age of around eleven to twelve years, it could be that a particular importance is attached to educating children about the importance of preserving Sark's unique environment, including its dark sky, before they leave for more light-polluted areas. This is underlined by head-teacher Sarah Cottle – 'educating the young of the island will be crucial [...] they will be [...] actively involved in promoting and educating others about it', and Mandy – 'I might say to the kids, right meet me here tonight at six o'clock and we'll go stargazing'.

When SAstroS have run their *Starfest* events they have hired a mobile inflatable planetarium in order to foster interest in astronomy amongst children. For some children

58 Direct quote from Mandy interview.

59 Sarah Cottle, in Owens, 'Sark Dark Sky Community: A Dark Sky Island'; direct quote from Mandy and Lydia interviews.

living on Sark it may take time before they begin to notice how special the night sky there can be as Mandy describes, 'Camping in the Seigneurie Gardens, being with a child who saw his first shooting star. He'd been here five years but how much time had he spent looking up?'[60]

Recalling my own childhood memories of the night sky, I was surprised at the stark difference between my experience and those of the participants: 'lots of times lying awake, insomniac, looking out at the night, hating the dark, wanting dawn to come'.[61] I wondered what the island's children felt if they woke at night. Perhaps my experience may be compared with the fear Ekirch conjectures prehistoric people may have had of the light not returning.[62]

FEAR AND FEARLESSNESS OF THE DARK

Fear of the dark was one of my concerns. For most participants this was usually associated with the possibility of meeting strangers, and newcomers quickly come to recognise many of the other inhabitants and feel safe. Several participants however described feeling occasionally 'unnerved' or 'freaked out' when out at night, and this seemed to be related to something unexpected occurring or their imagination playing tricks, as when Mandy describes, 'going past the cemetery, the wind blowing, you get peculiar

60 Direct quote from Mandy interview.

61 Direct quote from author's personal journal.

62 Ekirch, *At Day's Close*, p. 3.

noises'.[63] Such feelings may be associated with the symbolic role darkness has traditionally played as something evil.[64]

However, in contrast to Tim Edensor who suggests darkness may not always be appreciated, or indeed even feared, my participants who were actually brought up in rural areas or who have lived on Sark all their lives had no such concerns. For example, according to Gerry, 'as a child my grandparents lived in the country [...] no streetlights [...] I had no problem with the dark', whilst Roz explained, 'my kids, they've been brought up here [...] they'll go on their bikes ahead of me in pitch darkness'.[65] Moreover, the potential for darkness to promote positive transformation and psychological healing, as noted by Edensor and Bliss, is reinforced by Lydia: 'I've looked out at the dark and tried to kind of inhale the outside [...] get it into your body [...] there's a certain kind of comfort when you've had a bad day [...] powerful, transformative'; and Roz: 'changes my whole perception of things, clearing clouds'.[66] Jeremy had a different take on what darkness mean to him, 'actually daytime is the freaky bit, the natural state is for it to be dark, the rest of the universe is dark, all the space in between'.[67]

For several people being out in the dark was a sensual and aesthetic experience as demonstrated by Rosie,

63 Direct quote from Mandy interview.

64 Galinier *et al.*, *Anthropology of the Night*, p. 820.

65 Edensor, *Gloomy City*, p. 11, direct quotes from Gerry and Roz interviews.

66 Direct quote from Lydia and Roz interviews; Edensor, *Gloomy City*, p. 13; Bliss, 'In Praise of Sweet Darkness', p. 174.

67 Direct quote from Jeremy interview.

> taking all the artists out at night, I wanted to show them there's something about being completely in the dark [...] it heightens your other senses, you can hear the birds, a Manx shearwater calling from the beach, it sounded completely eerie [...] one of them (*the artists*) did a painting of the Milky Way here.[68]

Miranda stated, 'you begin to get used to shades of darkness [...] like looking at a colourful landscape at night, there's the same beauty in the sky but the colours are shades of black and silver'.[69] Lydia described the particular intensity of the darkness, 'a soupy blackness, almost claustrophobic', but she also described the quality of a bright, full moon night, 'the moon has been like looking at the sun, very exciting, exhilarating, very pure.[70] Alex mentioned noticing the colours of stars, 'the yellow sulphur moon, I love that', whilst Lydia noticed individual characteristics, 'the colours, the brightness, they're not just stars, they have little personalities'.[71] Unsurprisingly perhaps, for the size of the population, there are a large number of artists and photographers on Sark.

Several participants who had spent long periods of time in urban areas arrived on Sark with a level of fear but quickly became confident in the dark. Although Lydia had spent the first twenty three years of her life living in an

68 Direct quote from Rosie interview.
69 Direct quote from Miranda interview.
70 Direct quote from Lydia interview.
71 Direct quotes from Alex and Lydia interviews.

English village with limited outdoor lighting she was unprepared for the darkness on Sark, 'when I moved here I had to train myself not to be scared [...] now I feel 100% safe [...] it wasn't people or things I was afraid of, it was just so black'.[72] Stories were recounted of having friends visit who were surprised (and embarrassed!) at struggling with the darkness. Mandy laughed remembering her visitor, 'he was literally on his hands and knees to see if he could feel the path – pitch black', whilst Gerry's young friend who had joined the army was, according to Gerry, 'frightened to leave my house to walk in the dark'.

This echoes my own experience of arriving on Sark for the first time; a slow, unconfident cyclist getting left behind by friends having to grapple with my fears as I slowly inched forward in pitch darkness. Being out at night on Sark I found myself walking much more slowly than usual to avoid hazards, wary of what might be out there, although more fearful of meeting one of the island's black rats than a person. At the same time however, I was also gradually tuning into myself, becoming more thoughtful and reflective. Thankfully I noticed my fear gradually diminishing on subsequent visits. Getting to know many of the local community, if only by sight, also helped dispel fear. As Mandy relates, 'it's usually the human element that makes you feel unsafe and you haven't got that here [...] especially in winter, you don't see people you don't know'.[73] But even if you are a visitor Annie had this advice, 'if you pass someone in

72 Direct quote from Lydia interview.

73 Direct quote from Mandy interview.

the dark say good night, good evening and you normally get a response, you don't need to duck into a hedge!'[74] If astronomical tourism to Sark increases however, there will be a greater chance of meeting more 'strangers' during the winter.

Although David Kopel and Michael Loatman comment that some people fear increased crime or loss of privacy if particular types of lighting are given up, comments by most participants demonstrate no evidence of this. Mandy remembered being told as a child, 'if an area was dark then it was dangerous, when I was brought up that was the message. That doesn't apply here because it's such a small community'.[75] Alex felt safer on Sark and 'never scared here [...] different when it's an urban dark, I'm not keen on that'.[76] Like me Mandy also recalls growing up with a streetlight right outside her bedroom window, 'living where I lived there was no such thing as darkness'.[77] Sark has always had extremely limited outdoor lighting, a very low crime rate and neighbours are generally perceived as friendly.[78] Puffin's comment corroborates this. He told me that 'I feel safe when alone in the dark and great for elderly relatives to be able to go out unaccompanied to friends'.[79] In

74 Direct quote from Annie focus group.

75 Kopel and Loatman, *Dark Sky*, pp. 8–9.

76 Kopel and Loatman, *Dark Sky*, pp. 8–9.

77 Direct quote from Mandy and Alex interviews.

78 Summary of Sark perception survey results: http://www.gov.sark.gg/Downloads/Press_Releases/2012_Press_Releases/121203_Survey_Results-Island-Wide_Survey.pdf [Accessed 25 May 2014].

79 Direct quote from Puffin email.

addition, Kopel and Loatman's hypothesis that people fearful of the dark may not welcome preservation of dark skies to encourage tourism could not be verified in my study, as none of my participants were afraid of darkness, and everyone wanted more tourists.[80]

THE DARK SKY MOVEMENT AND ASTRONOMICAL TOURISM

Kevindran Govender has argued that astronomical tourism can bring socio-economic benefits particularly to isolated communities.[81] The majority of participants agree, and saw Sark's award of Dark Sky status as positive in that there was a belief it would lead to increased astronomical tourism, vital for a geographically isolated island heavily dependent on visitors. Mandy's comment is typical: 'we're a Dark Sky Island and it's inspiring people to come here [...] best star viewing is in winter [...] when we don't tend to have tourists so that would only be a good thing'.[82] Sark Tourism's website promotes the island as 'a perfect wedding destination', and Lydia has been photographer for a number of these weddings. One particular wedding sticks in her mind: 'a couple came from America for four days to get married [...] to look at the stars and the whole weekend

80 Kopel and Loatman, *Dark Sky*, pp. 8–9.
81 Govender, *Astronomy*.
82 Direct quote from Mandy interview.

was cloudy – they were both obsessed with the stars'.[83] It apparently did not spoil their enjoyment of the day, and Sark does have many other charms on offer, but it does illustrate the danger of dark sky places relying solely on clear skies. Three other participants and a number of other residents I met referred to this wedding and the capriciousness of the sky that weekend. The story has clearly gone down in island history!

However, despite the possibility of increased tourist revenues Rosie refers to there having been some opposition to applying for Dark Sky status: 'there's elements on the island who pooh-poohed dark skies – they say, "dark skies, dark ages", they think we're backward'.[84] Alex also recounted hearing others question the need for 'official' recognition from outside bodies, 'they said, the skies are there, why do we need to have the status [...] why do we need these researchers coming here to tell us we qualify?'[85] Outside involvement in relation to promoting Sark's dark skies has generally been welcomed however. SAstroS itself was founded as a direct result of the award of Dark Sky status, as Annie and Reg describe,

> it was suggested by them (*the* IDA) it would be useful to have a society [...] the IDA doesn't want to stick labels on places that are naturally dark anyway, it's looking for com-

83 Direct quote from Lydia interview.
84 Direct quote from Rosie interview.
85 Direct quote from Alex interview.

> munities like ours where with a small adjustment they can achieve what the really dark places have,

and, 'SAstroS is the very first one [astronomy society on Sark] but I think that's partly because we've always just looked up and didn't think we needed to do anything'.[86] As so often the case, the dark sky can often be taken for granted.

Puffin saw the award as an incentive to preserve the dark sky, regardless of whether it led to increased tourism or not: 'I think knowing about the sky, planets, stars, etc. would make people consider the wonder of it all and take care of the planet, their relationships, etc. [...] I wish I could live forever to appreciate this'.[87] As Robyn Eckersley said, one stream of environmentalism 'appeals to the enlightened self-interest of the human community'.[88]

Like many small island cultures, Sark faces the challenge of preserving its identity whilst maintaining its economy. There was an assumption amongst participants that as tourism centres on the island's uniqueness, its perceived simpler, more traditional way of life and unspoilt natural environment, any astronomical tourists would be respectful and understand the need to preserve Sark's dark skies. Mandy's remark illustrates this, 'they're the same people who come to appreciate the wild flowers [...] they appreciate that what makes the dark sky possible is

86 Direct quotes from Annie and Reg focus group.

87 Direct quote from Puffin email.

88 Eckersley, *Environmentalism*, pp. 36–38.

no pollution'.[89] There were a number of references to the high cost and lengthy journey to get to Sark and a belief this helped 'weed' out tourists who would not appreciate Sark as Rosie's comment highlights, 'there's no airstrip so you have to come by boat [...] it's not everyone's cup of tea'.[90] Lydia, on the other hand, felt the time taken was an important part of the Sark experience: 'you have to take your time to get here, that's part of the treat [...] but it's not always easy here, mud, wind, no boats'.[91] The overall feeling was that astro-tourism was welcome as long as visitor numbers were small, 'like the Folk Festival' and the new observatory was welcomed as 'another thing you can use to sell your island'.[92]

Although research conducted by the Dark Sky movement focuses on raising awareness of the effects of light pollution and outdoor lighting design, there appeared to be a long-standing understanding of these issues amongst all participants. Martin regretted, 'light pollution in the UK has become a sickening intrusion' and Puffin agreed, 'in cities light[s] are angled wrong and cause pollution'.[93] Nearby Guernsey was also regarded as suffering from light pollution as Annie told me: 'a SAstroS member in Guernsey texted me one night saying "I can see Orion and I can see his belt" but he couldn't see his sword even though Guernsey is

89 Direct quote from Mandy interview.
90 Direct quote from Rosie and Lydia interviews.
91 Direct quote from Rosie and Lydia interviews.
92 Direct quote from Jeremy interview.
93 Rich and Longcore, *Ecological Consequences*; direct quotes from Martin and Puffin emails.

only six miles away. That's how much difference even the tiniest bit of light pollution can make'.[94] Jeremy (perhaps mindful of his legislative and financial role on the island) took a more pragmatic view, 'electricity's very expensive' he said, adding that the 'reason we have dark skies is we're too mean to pay for electricity'.[95] Perhaps cutting back on outdoor lighting for many on Sark is an economic decision not just an altruistic one. This attention to saving energy was underlined by a member of the audience during my presentation at the Island Hall who pointed out that it was wasteful to have the lights on in the room when it was sunny outside. Embarrassed, I quickly had them turned off.

There does not seem to have been any research looking at the financial consequences of dark sky status, or the costs and benefits of accommodating astronomical tourists, but a minority of participants believed there could be costs involved. Whilst Alex speculated, 'people might have resented having to buy different lights', Jeremy focused on costs of astronomical equipment, 'problem is you need an observatory or at least a decent telescope'.[96] The educational aspects of astronomical tourism and the opportunities it offers visitors to learn about the importance of maintaining dark skies, as researched by Frederick Collison and Kevin Poe, wasn't a particular issue for my participants. During my visit in March 2014, the Sark Visitor Centre did however have a small exhibition on the island's Dark Sky

94 Direct quote from Annie interview.
95 Direct quote from Jeremy interview.
96 Direct quotes from Alex and Jeremy interviews.

application process which was running the following year in an expanded form. There was also an acknowledgement amongst most participants who were incomers to Sark that they now engaged more with astronomical activities in different ways. Mandy was considering returning to higher education, 'may be doing an astronomy degree [...] wouldn't have occurred to me to do if I wasn't living here', whereas Martin was content that living on Sark 'has rekindled my interest in astronomy'.[97] Even as occasional visitors to Sark, my partner and I found ourselves eager to know more about what we were seeing in the night sky and enrolled on an astronomy course in Edinburgh.

SKYSCAPE AND LANDSCAPE

For most residents, living on Sark involves living in close relationship with nature. Getting about involves walking or cycling in all types of weather. There are no high-rise apartments; everyone lives in a house and opens their front door to find the natural world on their doorstep. As Lydia said: 'here your animal senses come out, your hearing and sense of smell [...] nothing is tarmacked over, the earth is under your feet'.[98] Rosie described the dark sky as, 'all enveloping [...] part of everyday life' and forming a backdrop to their lives in the same way as do the many Sark plants and animals some of which are rarely found on the Brit-

97 Direct quote from Mandy interview and Martin email.

98 Direct quotes from Lydia interview.

ish mainland.[99] This is perhaps why the sky is widely seen by participants like Alex as part of nature. He said that 'the dark sky is integral, it's what Sark is, it's part of nature here'.[100] And when Rosie commented, 'it's important to keep that connection with nature', she was specifically referring to the connection she felt to the dark sky.[101] For me, there are close similarities between how participants expressed their understanding of the concept of an interaction between their own wellbeing and the larger ecosystem of the natural environment, which for them included the sky, and views articulated in more theoretical terms by authors such as David Abram, Freya Mathews and Theodore Roszak.[102]

Another important scholar, Tim Ingold explicitly discussed the concept of sky as part of landscape.[103] On Sark, on dark nights, one can experience the boundary between land and sky disappearing. Because there are no streetlights on the island, on moonless nights away from the pale roads the dark sky seems to remove the horizon making sky and land appear as one – a visual phenomenon we do not usually experience. Being outdoors at night on Sark means being *in* the sky and feeling connected to it, as Lydia describes, 'I am in it and feel enveloped, part of it', and Alex similarly experiences, 'I go out into it and feel immersed in

99 Direct quote from Rosie interview.
100 Direct quote from Alex interview.
101 Direct quote from Rosie interview.
102 Abram, *Spell*; Mathews, *Ecological Self*; Roszak *et al.*, *Ecopsychology*.
103 Ingold, *Being Alive*, p. 127.

it'.[104] Being *in* the sky can also be spatially disorientating as Rosie relates, 'you don't know where you are, no horizon, you can't place yourself'.[105] Personally, walking across a springy field one dense, dark night on Sark I also had the strange sensation that I was simultaneously both walking in the sky and also being supported by it. As Jack Borden remarked, we are in the sky rather than below it.[106] Also, after my presentation at the Island Hall, Annie recalled that when Bogard visited Sark he observed that the stars came down as far as the horizon, a phenomenon he had never noticed elsewhere.[107] For the people of Sark the sky is very much part of nature.

Previous studies - by Stephan Mayer and Cynthia Frantz, Nisbet *et al.* and Schultz - have investigated whether a greater sense of connectedness to nature results in a greater responsibility to protect it.[108] As Daniel Brown has found, being outdoors under dark skies encourages sustainable behaviour as regards light pollution.[109] Annie's response supports this claim:

> watching stars, meteor showers and comets [...] connects humanity to its environment [...] the animals and birds, what mankind does now with flooding the world with arti-

104 Direct quote from Alex and Lydia interviews.
105 Direct quote from Rosie interview.
106 Borden, *A New View*.
107 Author's personal conversation with Annie 5 Sep 2015.
108 Mayer and Frantz, *Connectedness*; Nisbet *et al.*, *Nature Relatedness*; Schultz, *Inclusion*, pp. 62–78.
109 Brown, *Higher Education*, p. 68.

> ficial lighting, it's creating eclipses for them [...] on a black rainy night, I've gone home with a torch and been careful just to shine it on the road because as I pass if there's a bird in the hedge it will wake up.[110]

In the same vein, comments by Rosie and Paula correlate with Moore's hypothesis that darkness gives an opportunity to cultivate a closer connection with the natural world, 'recently I've been setting a moth trap in the garden [...] seeing the bats', and 'I loved doing night carriages. One night I remember particularly there were big old trees at the side of the road and my little horse kept jumping over the bars of dark, the moon shadows'.[111] Paula, perhaps unlike some residents, enjoyed the rare occasions when she came across a black rat at night, 'the black rats, they're very pretty, they're out at night doing their things as nature intended'.[112]

The idea of the land-sky connection as represented by astronomical, lunar-based, agriculture was actually raised in the Q&A session following my presentation at the Island Hall, when someone asked whether biodynamic gardening was commonly practiced on Sark. This is a system of gardening that developed out of Rudolph Steiner's ideas of a unified approach to agriculture which one online source says, 'relates the ecology of the farm-organism to that of

110 Direct quote from Annie focus group.
111 Direct quote from Rosie interview and Paula focus group; Moore, *Gifts*, p. 12.
112 Direct quote from Paula focus group.

the entire cosmos'.[113] As one of its tenets is that the moon has an influence on plants, the questioner had wondered if gardeners on Sark were more likely to pay attention to moon phases when they planted. Speaking to Roz a few days later, I discovered that she and another gardener Helen had established a permaculture market garden at Stocks Hotel; permaculture is a system which also advocates planting by the moon.[114] The final entry in my own journal reads,

> so here is the Sark sky story. The night sky is not just seen but 'felt' by Sarkese and everyone else. It is in them and they are in it, regardless of whether it's intentionally sought out or come upon with soft fascination. It's widely talked about and children are encouraged to step into it from an early age. Everyone is welcome to come and share in it. This is because it's known that to connect with it strengthens a connection to our own inner cosmoses and helps preserve all of our histories.[115]

CONCLUSIONS

To conclude, it is clear that dark skies have many benefits, including a high level of enjoyment and value placed on ob-

113 Tom Petherick, Biodynamic Association, 'Getting started with Biodynamic Gardening', http://www.biodynamic.org.uk/farming-amp-gardening/getting-started-with-biodynamic-gardening/ [Accessed 25 Jan 2016].
114 'Planting by the Moon', https://permacultureprinciples.com/post/planting-by-the-moon/ [Accessed 4 Oct 2016].
115 Direct quote from author's personal journal.

serving the night sky with others, facilitating family/community connections and friendship, the transmission of sky stories to others, the evocation of childhood sky memories, fearlessness of the dark and a sense that sky and land can appear as one and a range of positive (and sometimes transformative) feelings.

CHAPTER 5:

IMAGES EMERGE: REFLECTIONS AND FUTURE PERSPECTIVES

REFLECTIONS

A MEMORY – my first trip to Sark, soaking in my friends' hot tub outside their house near Grand Greve, I saw a bright blinking light, moving just above where the horizon would normally be. But this was a dark night on Sark so the horizon was barely visible. I felt suspended in space. And above me, orbiting two hundred and twenty miles above the earth and incredibly bright, I was watching the International Space Station whizz by. I couldn't contain my amazement that this was a vehicle full of human beings. 'Oh, we've seen it loads of times', my friends reacted. That is when I first became intrigued by what it must be like to live somewhere where such a sight was commonplace. I realised that even if the Station flew right over Edinburgh, the plethora of streetlights would probably mean I would miss it anyway.

When I was studying the MA in Cultural Astronomy and Astrology, my fellow students and I were given a simple project: photograph the full moon on a particular night. As the MA is an online course and has students from every continent, our resulting photographs were all quite different. Our webinar discussion the following week focused on our global view of the same celestial object. Those of us in the northern hemisphere saw the moon apparently 'upside down' compared to the moon seen by those in the southern hemisphere and the colours we observed differed due to factors such as atmospheric pollution. Similarly, in 2007 to celebrate Scotland's Year of Highland Culture, pupils from seven schools in northern Scotland chose significant dates from Highland history and plotted them in light years from the earth. They then developed stories related to the dates in their local history and chose a star for their area from a selection of circumpolar stars, the stars visible all year round in that part of Scotland. As soon as each area had its own star, they were joined up to form the 'Highland Constellation'. A competition amongst the schools for a name resulted in it being called the 'Jumping Fish'. Both the story of the lunar photographs and story of the creation of a new constellation illustrate that whether our sky conversations take place internationally or more locally, we are all living on the same planet, seeing the same sky. Our conversations connect us not only with the sky but with each other.

As I began revising my MA dissertation for this book I found myself thinking more about my motivation for doing the project in the first place. I have come to realise that the roots of the project predate my first visit to Sark. Rewind the clock fifty years and my five year old self is already a confirmed bookworm, devouring the books in my local library. Every so often I would imagine writing my own book and practiced doing my autograph, preparing for a book signing. Next to my name I always drew a symbol, a circle with an arrow going through the heart of it pointing to the right. At the time I had no idea what it meant. The circle and the arrow are ancient symbols that have various meanings attached to them. Reflecting on my life, and the various academic and spiritual journeys I have made, I can now see that my circle/arrow represents the process of connecting up the part of me that needs to be grounded with the part that enjoys having my head in the clouds and seeing those 'holes in the sky'.

Although dark night skies can help us better understand and make sense of life here on earth, in contrast to my experiences on Sark, I rarely have conversations about the sky with anyone in Edinburgh. Time spent with friends or family generally involves being indoors – the occasions we are outside are usually during daylight hours. Opportunities to speak about the night sky are rare; the sky simply does not feature in our lives in the way it does for people on Sark. And yet sharing experiences of meteors, comets and eclipses – events that are outside of everyday lives – connects us more powerfully with 'the bigger picture' whilst anchoring us to people or places we hold dear, whether on Sark or elsewhere. In August 1999 I travelled to Dartmoor with my husband and son, staying with his parents on the way, to watch the total solar eclipse. My late mother-in-law's dire warnings about going blind are still very fresh in my memory! The story we all still recall however is sitting on the moor feeling both excited and scared as the moon gradually began to cover the sun, noticing that all the birds had stopped singing, and that the scattered groups of Dartmoor ponies had formed a long line and were disappearing over the brow of the hill. It is one of my fondest family memories. When David Bowie died in early 2016, the Belgian radio station Studio Brussel and the MIRA Public Observatory paid tribute by proposing a new asterism for him. The gesture was a symbolic one (rather than an official astronomical update), with the seven stars in the asterism forming a lightning bolt, a motif Bowie often used. These days our sky stories may not tell of gods and goddesses or beasts, as our ancestors' stories did, but they are still sky stories. They help build family and community connections, and they emphasise the similarity of our experiences, not just the differences. Perhaps this is especially important in a world where single-person households are on the increase and many communities have become fragmented.

Sark is clearly ahead in terms of accessing beautiful night skies and has a small population who have a strong desire to preserve them. Many people there enjoy being out at night looking at the sky, and for them it has a similar positive impact on their wellbeing as walking amongst the Sark wild flowers or watching bottlenose dolphins from the cliff tops. Although many Dark Sky places are in rural areas I see no reason why the type of community-driven initiative that took place on Sark cannot be at least attempted in locales within urban areas. There is evidence showing how urban green spaces can help bring people together and create community cohesion. Perhaps this model can be extended to our urban skies.

However there can often be a tension, even for some on Sark, between a desire to see the stars, which many nostalgically recall from childhood, and a desire to feel safe at night. Darkness is still associated with crime, and any changes local authorities make to outdoor lighting policies are often opposed. Also, as we try to cram more activities into each day, blurring the boundary between work and leisure, day and night, we expect services such as supermarkets, garages and restaurants to be open later. More than twenty years ago a friend of mine moved to a mobile home site on the outskirts of Edinburgh with beautiful views of the nearby Pentland Hills and the sky above. These days the lights from the new twenty-four-hour supermarket built across the road mean the sky is no longer ever very dark. All over the world lights are being left on for longer. Although artificial light and information and communications technology have given us the opportunity to work twenty four hours a day, our circadian rhythms or 'biological clock', which help ensure our sleeping and digestion functions properly at the right time of day, have not changed. Unsurprisingly sleep and other health problems are on the increase – our bodies need the dark. Yet by avoiding the darkness, what are we trying to tame or keep at bay? Is it our 'inner wildness' as Jung believed? Perhaps with all our technological advances we would rather not remember that we are still animals with animal needs for food and sleep.

Artificial lights also litter the skyscape. On Sark, Mandy and Gerry recalled enjoying being out at night with others counting satellites but 'Jay' was worried about 'the rubbish circling the earth'.[1] With increasing demand for satellite TV, broadband and telephony, the number of satellites they are likely to see will continue to rise. In the summer of 2016 Moscow State University of Mechanical Engineering's *Mayak* (beacon) project hopes to launch a reflector satellite which is being described as an 'artificial star'. Costs for the project were crowdfunded and the funding target quickly reached. The satellite will illuminate different locations on the Earth by reflecting sunlight as it rotates, becoming the brightest object in the sky other than the sun. Apparently the satellite will serve no particular purpose, other than proving that it can be done...

Sometimes planes can be mistaken for stars before the eye adjusts to their movement. As the market across the world for air travel continues to grow there will be many more aircraft in the sky. Whilst there are concerns about aviation causing noise pollution and negative effects on air quality, aircraft lights are also contributing to obscuring views of the night sky. The IDA concentrates much of its efforts into developing a conservation programme for Dark Sky places and reducing light pollution. Perhaps attention also needs to now focus on how to best avoid cluttering up the sky with bright lights. Having a no fly zone like Sark has is not feasible for many areas and only addresses the problem of aircraft lights. It is not too hard to imagine a time in

1 Direct quote from 'Jay' email.

the future where the 'holes in the sky' are not the stars but planes, satellites, and other space hardware.

My project involved a small island which is unusual in many ways, and I wonder if the relationship many Sark residents have with their dark sky is transferrable to other places. As I was writing this final chapter I read that Moffat, a small village in southern Scotland with around 2,500 residents, had just been awarded Dark Sky Community status, with the hope being this will lead to increased tourism to the area. The process took three years and involved the local community council successfully lobbying Dumfries and Galloway Council to make the necessary changes to their outdoor lighting policies. Moffat has cars and street lights and is not a remote place, it is just over fifty miles from Edinburgh, yet the community's achievement shows what can be done by motivated communities on the mainland.

POSSIBLE AREAS FOR FUTURE RESEARCH

My research focused on adults' experiences of the night sky, but children's experiences would also be worth exploring particularly as childhood sky memories featured prominently in the research. In addition, although I did not interview any older people who had lived on Sark all their lives (other than those who were SAstroS members), given this group were often referred to as, 'taking the sky for granted', their views might also deserve further study. As astronomical tourism to Sark and other Dark Sky Places grows, the costs of accommodating the needs of such tourists and their motivation to then implement changes to lighting at their homes and engage more with astronomical activities could also be investigated. One of the areas the Glasgow University Galloway project will look at is the apparent rise in popularity of dark sky places. What motivates people to visit these places? As the number of places receiving dark sky status continues to grow, it is as yet unknown whether having so many may lead to dark sky places being viewed as a less 'exclusive' visitor experience. What facilities will visitors expect, what will get them to return again? However Sark, unlike some other places, has the island itself as a unique resource, not only its dark sky.

The method I chose to do my research – intuitive inquiry, with its focus on reaching a deeper understanding of how people on Sark felt about their dark sky – could perhaps be applied to different types of dark sky places. Comparisons could then be made between the experience of living in a Dark Sky Community as opposed to, for example, a Dark Sky Development of Distinction or Reserve. Finally, the unusual sensation of feeling one is *in* the sky when out on a dark night could also be further explored within other Dark Sky Communities.

FINAL THOUGHTS

The number of participants in my research was relatively small, but I believe the project and its findings are important because the same feelings in relation to the dark sky were described repeatedly. My research therefore has begun to address the missing sky factor within the fields of ecopsychology and health and environmental psychology, connecting up land and sky. My hope is that these findings can be used to strengthen the Dark Skies movement's claims that dark night skies can have a positive impact on wellbeing. Hearing of the research, the IDA asked me to write a blog post, 'The Psychology of Dark Skies'.[2] The post received almost 300 'likes' on the IDA's Facebook page, and subsequently I have been approached by a number of journalists interested in my research. Also, someone living in a very heavily light-polluted city in the US contacted me to ask if I would consider offering her online counselling. She has been suffering from poor mental health, which she attributes to the lack of access to the stars. Clearly there is growing interest in this field!

Conducting research often facilitates transformation of the researcher and leads to, 'important, meaningful and sometimes profound changes in one's attitudes of one's views of oneself and of the world at large'.[3] During the course of my research I have become much more aware of the increasing light pollution in my home city of Edinburgh. I have begun engaging in attempts to influence both my local authorities' and workplaces' policy decisions regarding outdoor lighting – challenging assumptions about what lights need to be on all night and suggesting partial or dimmer lighting whenever possible. The United Nations

2 Ada Blair, IDA, 'The Psychology of Dark Skies', http://darksky.org/the-psychology-of-dark-skies/ [Accessed 26 Jan 2016].

3 Anderson and Braud, *Transforming Self and Others*, p. xv.

General Assembly designated 2015, 'International Year of Light and Light-based Technologies' (IYL), recognising the vital role light plays in our lives particularly with regard to promoting global solutions in many fields including sustainable development, energy, education and health.[4] Whilst, ironically, it was hoped that one of the outcomes of IYL might be a greater awareness of darkness, I find myself wondering how much more could be achieved for our wellbeing if we were to similarly celebrate the dark and proclaim an 'International Year of Darkness'. Also, as a therapist who has in the past occasionally conducted therapy sessions outdoors in daytime, I am now considering the therapeutic benefits of having sessions under a dark night sky.

As to my need to have my feet on the ground and my head in the clouds, back home in Edinburgh I remember that feeling of being held in the sky on a dark night on Sark. Looking up at the dark sky, feeling awed and small, I get a sense of our island Earth's own vulnerability in the vastness of all those other galaxies. With no sense of where the horizon is, no differentiation between land and sky, I feel I have lost my footing but at the same time feel no fear, totally at ease and supported, suspended in a union between heaven and earth.

4 International Year of Light and Light-based Technologies, IYL2015, http://www.light2015.org/Home.html [Accessed 22 Jan 2016].

Appendix

LIST OF INTERVIEW QUESTIONS

- How long have you lived on Sark?
- Have you lived in other places or do you regularly travel to other places?
- What's your earliest memory of looking at the night sky on Sark? Describe this in as much detail as you can.
- How often do you find yourself noticing the night sky?
- Can you recognise any constellations, planets, etc.?
- Have you heard any stories about the night sky?
- What feelings come up for you when you look at the night sky? Describe these feelings in as much detail as you can.
- What's the most powerful experience you've had when looking at the night sky? How did that experience make you feel at the time?

- Do you think the experience has had any longer-term impacts for you? On physical, emotional, spiritual wellbeing?

- Do you consider yourself to be a spiritual and/or religious person?

- Do you feel there are any benefits to watching the night sky? If so, what benefits have you experienced yourself?

- (*If you spend time away regularly from Sark*) Are you aware of any differences in how the night sky looks elsewhere as compared to Sark?

- Is how you feel about the night sky elsewhere different to how you feel about the night sky on Sark?

- Do you ever have conversations with others about the Sark night sky? If so, what experiences, etc. have others reported?

- How do you feel about being out at night in the dark, on Sark and elsewhere?

Bibliography

MANUSCRIPT/UNPUBLISHED SOURCES

Brown, Daniel. *How Can Higher Education Support Education for Sustainable Development? What Can Critical Place-Based Learning Offer?* (Unpublished MS, University of Nottingham, 2013).

Holbrook, Jarita. *Sky Knowledge, Celestial Names, and Light Pollution* (Unpublished MS, University of Arizona, 2009).

PRIMARY SOURCES

Andrews, Chris, Zoller, Renate and McKee, Amy eds. *Art for the Love of Sark*. Oxford: Gateway Publishing Ltd., 2012.

BBC News. 'Barclay brothers' Sark hotels to close', http://www.bbc.co.uk/news/world-europe-guernsey-30035969.

———. 'Sark, the First Dark Sky Island, Gets Observatory', http://www.bbc.co.uk/news/world-europe-guernsey-32596199.

———. 'Sark's Astronomical Observatory Opens', http://www.bbc.co.uk/news/world-europe-guernsey-34495607.

BBC Two. 'An Island Parish', http://www.bbc.co.uk/programmes/b006t6m6.

Blair, Ada. *Personal journal*.

Bliss, Shepherd. 'In Praise of Sweet Darkness', in Buzzell and Chalquist, *Ecotherapy: Healing with Nature in Mind*.

Bogard, Paul, ed. *Let There Be Night: Testimony on Behalf of the Dark*. Nevada: University of Nevada Press, 2008.

Branch, Michael P., 'Ladder to the Pleiades', in Bogard, *Let There Be Night*.

Burke, Edmund, 'On the Sublime and Beautiful', https://ebooks.adelaide.edu.au/b/burke/edmund/sublime/part4.html

Cottle, Sarah, in Owens, 'Sark Dark Sky Community: A Dark Sky Island'.

Guille, Rosanne. *Sark Sketchbook: Journal of a Local Artist*. Sark: Cat Rock Publications, 2004.

Harris, Sara. 'Beyond the "Big Lie": How One Therapist Began to Wake Up', in Buzzell and Chalquist, *Ecotherapy: Healing with Nature in Mind*.

Hugo, Victor. 'Toilers of the Sea', http://www.online-literature.com/victor_hugo/toilers-of-the-sea/2/

International Dark-Sky Association. 'About IDA', http://www.darksky.org/about-ida

James, Julie. 'Fifth International Dark Sky Reserve Designated in Wales', http://www.darksky.org/night-sky-conservation/283

James, William. 'On a Certain Blindness in Human Beings', http://books.google.com/books?isbn=0141956585

Jung, C. G. *Flying Saucers: A Modern Myth of Things Seen in the Sky*. Princeton, NJ: Princeton University Press, 1979.

———. *Jung Speaking: Interviews and Encounters*. William McGuire and R. F. C. Hull, eds. Princeton, NJ: Princeton University Press, 1977.

———. *Man and his symbols*. New York: Random House, 1964.

———. *The Structure and Dynamics of the Psyche*. *The Collected Works* vol. 8, 2nd ed. Translated by R. F. C. Hull. Princeton, NJ: Princeton University Press, 1969.

Lomax, John A. and Alan Lomax. *American Ballads and Folk Songs*. New York: Dover Publications, 1994.

M magazine, 'Interview: Emma Pollock', http://www.m-magazine.co.uk/features/interviews/interview-emma-pollock/

MacNeice, Louis. *Holes in the Sky: Poems, 1944–1947*. London: Faber and Faber, 1948.

Moore, Kathleen Dean. 'The gifts of darkness', in Bogard, *Let There Be Night*.

Outdoor Nation, 'Q&A with Annie Daschinger and Jo Birch: The World's First 'Dark Sky' Island', http://blog.outdoornation.org/qa-with-annie-dachinger-and-jo-birch-the-worlds-first-dark-sky-island/

Owens, Steve. 'Sark Dark Sky Community. A Dark Sky Island. Application to the International Dark-skies Association', http://www.ida.darksky.org/assets/documents/dark%20sky%20community%20application.pdf

Peake, Mervyn. *Collected Poems*. Manchester: Carcanet Press, 2008.

———. *Mr. Pye*. London: Vintage Books, 1999.

Philip's Astronomy. *Philip's Dark Skies Map Britain and Ireland: Darkest Observing Sites in the British Isles*. London: Philip's Astronomy, 2004.

Robertson, Christina. 'Circadian Heart', in Bogard, *Let There Be Night*.

Royal Astronomical Society. '£500k for Public Engagement in Astronomy and Geophysics: Six Teams Win RAS Funding', http://www.ras.org.uk/news-and-press/news-archive/259-news-2015/2629-460k-for-public-engagement-in-astronomy-and-geophysics-six-teams-win-ras-funding

———. 'Sark becomes the world's first dark sky island', http://www.ras.org.uk/news-and-press/217-news2011/1920-sark-becomes-worlds-first-dark-sky-island

Ruskin, John. 'Of the Open Sky, Modern Painters I, Part II, Section III', http://www.lancaster.ac.uk/depts/ruskinlib/Modern%20Painters

Sark Chief Pleas. 'A Vision for Sark', www.gov.sark.gg/Downloads/Reports/A_Vision_for_Sark.pdf.

———. 'Sark Island-Wide Opinion Survey SWOT Analysis', http://www.gov.sark.gg/Downloads/Reports/SWOT_Analysis_290912.pdf

———. 'Summary of Sark Perception Survey Results', http://www.gov.sark.gg/Downloads/Press_Releases/2012_Press_Releases/121203_Survey_Results-Island-Wide_Survey.pdf

'Sark Hailed as the World's First Day Sky Island', http://www.socsercq.sark.gg/News%20and%20Projects/darkskiespressrelease.html

Sark Tourism. 'Star Gazing', http://www.sark.co.uk/star-gazing-12757/

———. 'What eclipse?' http://www.sark.co.uk/what-eclipse-9047/

'Sercq – La Valette Campsite', http://www.sercq.com/la_Valette_Campsite.html

Starlight Initiative. 'Objectives of the Starlight Initiative' http://www.starlight2007.net/index.php?option=com_content&view=article&id=199&Itemid=81&lang=en

Tate. 'Joseph Mallord William Turner, "La Coupée, Sark Island", ?1832', http://www.tate.org.uk/art/artworks/turner-la-coupee-sark-island-d23637

Tonks, Peter. 'Sark Dark Sky Community: A Dark Sky Island', http://www.darksky.org/assets/documents/dark%20sky%20community%20application.pdf

La Trobe, G. and L. *La Trobe Guide to the Coasts, Caves and Bays of Sark.* Seventh revised edition, eds. Jeremy Latrobe-Bateman and Rob Pilsworth. Sark: Lazarus Publications NFP, 2014.

SECONDARY SOURCES

Abram, David. *The Spell of the Sensuous: Perception and Language in the More-Than-Human World*. New York: Vintage Books, 1996.

Anderson, Rosemarie. 'An Epistemology of the Heart for Scientific Inquiry'. *The Humanistic Psychologist* 32.4 (2004): pp. 307–41.

———. 'Intuitive Inquiry: Interpreting Objective and Subjective Data'. *ReVision: Journal of Consciousness and Transformation* 22.4 (2004): pp. 31–39.

———. 'Thematic Content Analysis: Descriptive Presentation of Qualitative Data', http://www.wellknowingconsulting.org/publications/pdfs/ThematicContentAnalysis.pdf 2007.

——— and Braud, William. *Transforming Self and Others through Research: Transpersonal Research Methods and Skills for the Human Sciences and Humanities*. New York: SUNY Press, 2011.

Atkins, Stephen, Husain, Sohail and Storey, Angele. *The Influence of Street Lighting on Crime and Fear of Crime. Crime Prevention Unit Paper No. 28*. London: Home Office, 1991.

Atkinson, Rowland and Flint. 'Snowball Sampling', http://srmo.sagepub.com/view/the-sage-encyclopedia-of-social-science-research-methods/n931.xml.

Aveni, Anthony, *Conversing with the Planets: How Science and Myth Invented the Cosmos*. Boulder, CO: University Press of Colorado, 2002.

———. *People and the Sky: Our Ancestors and the Cosmos*. London: Thames and Hudson, 2008.

Barrow, John D., *The Artful Universe*. Oxford: Clarendon Press, 1995.

Bell, Julia. 'Why Writers Treasure Islands: Isolated, Remote, Defended – They're Great Places for Story-Telling', http://www.independent.co.uk/arts-entertainment/books/features/why-writers-treasure-islands-isolated-remote-defended-theyre-great-places-for-story-telling-10391558.html

Biernacki, Patrick and Dan Waldorf. 'Snowball Sampling Problems and Techniques of Chain Referral Sampling', *Sociological Methods and Research* 10.2 (1981): pp.141–63.

Blackburn, Michele, Curtis Burney and Louis Fisher. *Management of Hatchling Misorientation on Urban Beaches of Broward County, Florida: Effects of Lighting Ordinances and Decreased Nest Relocation. For the Broward County Board of County Commissioners, Environmental Protection Department* (2007).

Blair, Ada. 'The Psychology of Dark Skies', http://darksky.org/the-psychology-of-dark-skies/

Blankenship, Diane C. *Applied Research and Evaluation Methods in Recreation*. Champaign, IL: Human Kinetics, 2010.

Blask, David, George Brainard, Ronald Gibbons, Steven Lockley, Richard Stevens, and Mario Motta. 'Adverse Health Effects of Nighttime Lighting. Comments on American Medical Association Policy Statement', *American Journal of Preventive Medicine* 45.3 (2013): pp. 343–46.

Bloor, Michael, Jane Frankland, Michele Thomas and Kate Robson. *Focus Groups in Social Research*. London: SAGE, 2001.

Bogard, Paul, 'Let there be dark', *Los Angeles Times*, http://articles.latimes.com/2012/dec/21/opinion/la-oe-bogard-night-sky-20121221

———. *The End of Night: Searching for Natural Darkness in an Age of Artificial Light*. London: Fourth Estate, 2013.

Borden, Jack. 'For a New View of the World: Sky Walk', http://www.prevention.com/fitness/fitness-tips/reduce-stress-sky-walking

Boyatzis, R. E. *Transforming Qualitative Information: Thematic Analysis and Code Development*. London: SAGE Publications, 1998.

Braud, William. 'Towards a More Satisfying and Effective Form of Research', http://contemporarypsychotherapy.org/vol-2-no-1/towards-a-more-satisfying-and-effective-form-of-research/

——— and Anderson, Rosemarie. *Transpersonal Research Methods for the Social Sciences*. Thousand Oaks, CA: Sage Publications, 1998.

Buzzell, Linda and Chalquist, Craig, eds. *Ecotherapy: Healing with Nature in Mind*. San Francisco: Sierra Club Books, 2009.

Campion, Nicholas. *A History of Western Astrology. Volume 1: The Ancient and Classical Worlds*. London: Continuum Books, 2008.

———. *Astrology and Cosmology in the World's Religions*. New York: New York University Press, 2012.

Chepesiuk, Ron. 'Missing the dark', *Environmental Health Perspectives* 117.1 (2009): A20-27.

Cinzano, P. Falchi, F. and Elvidge, C. D. 'The First World Atlas of Artificial Night Sky Brightness', *Monthly Notices of the Royal Astronomical Society* 328.3 (2001): pp. 689–707.

Coleman, Mark. *Awake in the Wild: Mindfulness in Nature as Path to Self-Discovery*. San Francisco: Inner Ocean Publishing, 2006.

Collison, Frederick M. and Poe, Kevin. 'Astronomical Tourism: The Astronomy and Dark Sky Program at Bryce Canyon National Park', *Tourism Management Perspectives* 7 (2013): pp. 1–15.

Connell, J. 'Island Dreaming: The Contemplation of Polynesian Paradise', *Journal of Historical Geography* 29.4 (2003): pp. 554–82.

Coysh, Victor. *Sark: The Last Stronghold of Feudalism*. Guernsey: Toucan Press, 1982.

CPRE. 'Lighting nuisance survey 2009/10: Report', http://www.cpre.org.uk/resources/countryside/dark-skies/item/1974-lighting-nuisance-survey-2009-10-report

———. 'Star Count 2014: a dark outlook for starry skies', http://www.cpre.org.uk/media-centre/latest-news-releases/item/3583-star-count-2014-a-dark-outlook-for-starry-skies

Cresswell, John W. *Qualitative Inquiry and Research Design: Choosing among Five Approaches*. Thousand Oaks, CA: Sage Publications, 2010.

———. *Research Design: Qualitative, Quantitative and Mixed Methods Approaches*. Second edition. London: Sage Publications, 2003.

Daniel, Glyn. 'The Forgotten Milestones and Blind Alleys of the Past'. RAIN (Royal Anthropological Institute of Great Britain and Northern Ireland) 33 (1979): pp. 3–6.

Dann, G. M. S. 'Tourism: The Nostalgia Industry of the Future' in *Global Tourism: The Next Decade*, ed. W. F. Theobald (Oxford: Butterworth-Heinemann Ltd., 1994), pp. 55–67.

Dark Skies Awareness. 'Light Pollution – What is it and Why is it Important to Know?' http://www.darkskiesawareness.org/faq-what-is-lp.php

Davis, F. *Yearning for Yesterday: A Sociology of Nostalgia*. New York: Free Press, 1979.

Davis, John. 'The Transpersonal Dimensions of Ecopsychology: Nature, Non-duality and Spiritual Practice', *The Humanistic Psychologist* 26.1–3 (1998): pp. 60–100.

Dodge, Rachel, Annette P. Daly, Jan Huyton and Lalage D. Sanderset. 'The Challenge of Defining Wellbeing', *International Journal of Wellbeing* 2.3 (2012): pp. 222–35.

Dunnett, Oliver. 'Contested Landscapes the Moral Geographies of Light Pollution in Britain', *Cultural Geographies* 22.4 (2015): pp. 619–36.

Eckersley, Robyn. *Environmentalism and Political Theory. Towards and Ecocentric Approach*. London: UCL Press Limited, 1992.

Edensor, Tim. 'The Gloomy City: Rethinking the Relationship between Light and Dark', http://usj.sagepub.com/content/early/2013/09/24/0042098013504009.full

———. 'Reconnecting with Darkness: Gloomy Landscapes, Lightless Places', *Social & Cultural Geography* 14.4 (2013): pp. 446–65.

Ekirch, A. Roger. 'At Day's Close: Night in Times Past', http://www.amazon.com/At-Days-Close-Night-Times-ebook/dp/B007HXFT2C.

EU Sky Route. 'European Astro Tourism Route', http://www.euskyroute.eu/european-astrotourism-route/

———. 'EU Sky Route Newsletter', http://www.euskyroute.eu/wp-content/uploads/2015/04/eu-sky-route-newsletter-160315_site.pdf.

Evernden, N. *The Social Creation of Nature*. Baltimore: Johns Hopkins University Press, 1992.

Fayos-Solá, Eduardo, Cipriano Marin and Jafar Jafari. 'Astrotourism: No Requiem for Meaningful Travel', PASOS: *Revista de Turismo y Patrimonio Cultural* 12.4 (2014): pp. 663–d71.

Ferrer, Jorge N. *Revisioning Transpersonal Theory: A Participatory Approach to Human Spirituality*. New York: SUNY Press, 2002.

Fox, R. 'Enhancing Spiritual Experience in Adventure Programs', in J. Miles and S. Priest, eds., *Adventure Programming*. State College, PA: Venture Publishing, 1999.

Fraknot, Andrew. 'Light Pollution', http://www.pbs.org/seeinginthedark/astronomy-topics/light-pollution.html.

Franz, Hölker, Timothy Moss, Barbara Griefahn, Werner Kloas, Christian C. Voigt, Dietrich Henckel, Andreas Hänel, Peter M. Kappeler, Stephan Völker, Axel Schwope, Steffen Franke, Dirk Uhrlandt, Jürgen Fischer, Reinhard Klenke, Christian Wolter, and Klement Tockner. 'The Dark Side of Light: A Transdisciplinary Research Agenda for Light Pollution Policy', *Ecology and Society* 15.4 (2010): A13.

Galinier, J., A. Becquelin, G. Bordin, L. Fontaine, A. Monod, F. Fourmaux, J. Roullet Ponce, P. Salzarulo, P. Simonnot, M. Therrien and I. Zilli. 'Anthropology of the Night: Cross-Disciplinary Investigations', *Current Anthropology* 51.6 (2010): pp. 819–47.

Globe at Night. 'About Globe at Night', http://www.globeatnight.org/about.php

———. 'What is light pollution?' http://www.globeatnight.org/light-pollution.php

Govender, Kevindran. 'Astronomy Can Foster Development', http://www.scidev.net/global/opinion/astronomy-can-foster-development-1.html

Guynup, Sharon. 'Light Pollution Taking Toll on Wildlife, Eco-Groups Say, National Geographic News', news.nationalgeographic.com/.../04/0417_030417_tvlightpollution.html

The Guardian. 'Bright Future for "Dark Sky" Sites as Astrotourism Grows in Appeal', http://www.theguardian.com/science/2015/apr/12/dark-sky-tourism-northumberland-kielder-observatory-northern-lights

Gwiazdzinski, Luc. 'La nuit dernière frontière Night – The Last Frontier', *Les Annales de la Recherche Urbaine, Plan Urbanisme – Construction – Architecture* 87 (2000): pp. 81–89.

Harper, Stephen. 'The Way of Wilderness', in Roszak *et al.*, *Ecopsychology. Restoring the Earth. Healing the Mind.*

HarperCollins. 'Bogard, Paul', http://www.harpercollins.co.uk/authors/11241/paul-bogard

Hartig, Terry and Evans, Gary W. 'Psychological Foundations of Nature Experience', in T. Garling and R. G. Golledge, eds., *Behavior and Environment: Psychological and Geographical Approaches*. Amsterdam: Elsevier/North Holland, 1993.

Hartig, Terry, Marlis Mang and Gary W. Evans. 'Restorative Effects of Natural Environment Experiences', *Environment and Behavior* 23.1 (1991): pp. 3–26.

Harvard Health Publications. 'Blue Light has a Dark Side', http://www.health.harvard.edu/staying-healthy/blue-light-has-a-dark-side

Hertz, D. G. 'Trauma and Nostalgia: New Aspects of the Coping of Aging Holocaust Survivors', *The Israel Journal of Psychiatry and Related Sciences* 27.4 (1990): pp. 189–98.

Hewison, R. *The Heritage Industry: Britain in a Climate of Decline*. London: Methuen London Ltd., 1987.

Historic England. 'Heritage and the Economy', p. 1, https://content.historicengland.org.uk/images-books/publications/heritage-and-the-economy/heritage-and-the-economy-2015.pdf

Hofer, J. 'Medical dissertation on nostalgia'. Translated by C. K. Anspach. *Bulletin of the History of Medicine* 2, pp. 376–91.

Holbrook, M. B. 'Nostalgia and Consumption Preferences: Some Emerging Patterns of Consumer Tastes'. *Journal of Consumer Research* 20.2 (1993): pp. 245–56.

———. 'Nostalgia Proneness and Consumer Tastes' in J. A. Howard ed., *Buyer Behaviour in Marketing Strategy*. Second edition. (Englewood Cliffs, NJ: Prentice-Hall, 1994), pp. 348–64.

Hunter, Tim and Crawford, David. 'The Economics of Light Pollution', ASP *Conference Series* 7 (1991): p. 99.

Ingold, Tim. *Being Alive. Essays on Movement, Knowledge and Description*. London and New York: Routledge, 2011.

———. *The Perception of the Environment. Essays on Livelihood, Dwelling and Skill*. London and New York: Routledge, 2000.

Institute for Tourism. 'What is Sustainable Tourism?', http://www.iztzg.hr/en/odrzivi_razvoj/sustainable_tourism/

International Dark- Sky Association. '5 Popular Myths About LED Streetlights', http://darksky.org/5-popular-myths-about-led-streetlights/

———. 'Lighting and Crime. Information Sheet No. 51', http://www.darksky.org/assets/documents/is051.pdf

'International Year of Light and Light-based Technologies, IYL2015, 2015', http://www.light2015.org/Home.html.

Johnson, Henry. 'Sark and Brecqhou Space, Politics and Power', *Shima: The International Journal of Research into Island Cultures* 8.1 (2014): pp. 9–33.

Kahn, Peter and Stephen Kellert. *Children and Nature: Psychological, Sociocultural, and Evolutionary Investigations*. Cambridge, MA: MIT Press, 2002.

Kaplan, Rachel. 'The Nature of the View from Home: Psychological Benefits'. *Environment and Behavior* 33.4 (2001): pp. 507–42.

——— and Stephen Kaplan. *The Experience of Nature: A Psychological Perspective*. Cambridge: Cambridge University Press, 1989.

Kaplan, Stephen. 'Human Nature and Environmentally Responsible Behavior', *Journal of Social Issues* 56.3 (2000): pp. 491–508.

———. 'The Restorative Benefits of Nature: Towards an Integrative Framework', *Journal of Environmental Psychology* 16 (1995): pp. 169–82.

Kelly, William E. 'Development of an Instrument to Measure Noctcaelador: Psychological Attachment to the Night-Sky'. *College Student Journal* 38.1 (2004): pp. 100–2.

———. 'Night-Sky Watching Attitudes Among College Students: A Preliminary Investigation', *College Student Journal* 37.2 (2003): pp. 194–96.

——— and Jason Batey, 'Criterion-Group Validity of the Noctcaelador Inventory: Differences between Astronomical Society Members and Controls', *Individual Differences Research* 3.3 (2005): pp. 200–3.

——— and Kathryn E. Kelly, 'Further Identification of Noctcaelador: An Underlying Factor Influencing Night-Sky Watching Behaviors', *Psychology and Education: An Interdisciplinary Journal* 40.3-4 (2003): pp. 26–27.

Kibby, M. 'Tourists on the Mother Road and the Information Superhighway', in *Reflection on International Tourism: Expressions of Culture, Identity, and Meaning in Tourism*, eds. M. Robinson, P. Long, N. Evans, R. Sharpley and J. Swarbrooke (Newcastle: University of Northumbria), pp. 139–49.

Kopel, David B. and Michael Loatman. 'Dark Sky Ordinances: How to Separate the Light from the Darkness'. Colorado: Independence Institute, 2006. http://www.davekopel.com/env/DarkSkies.pdf.

Koslovsky, Craig. *Evening's Empire: A History of the Night*. Cambridge: Cambridge University Press, 2011.

Krupp, Edwin C. *Echoes of the Ancient Skies: The Astronomy of Lost Civilizations*. New York: Dover Publications, 2003.

Louv, Richard. *Last Child in the Woods: Saving Our Children from Nature-Deficit Disorder*. Chapel Hill, NC: Algonquin Books of Chapel Hill, 2005.

Macnaghten, P. and J. Urry. *Contested Natures*. London: SAGE Publications, 1998.

McKinn, R. *The History of the BAA: The First Fifty Years*. London: BAA, 1990.

Marin, Cipriano and Jafar Jafari eds. *StarLight: Declaration in Defence of the Night Sky and the Right to Starlight (La Palma Declaration), International Conference in Defence of the Quality of the Night Sky and the Right to Observe the Stars, La Palma, Canary Islands, Spain, (April 19–20 2007)*. La Palma: Starlight Initiative, 2007.

Mathews, Freya. *The Ecological Self*. London: Routledge, 1991.

Mausner, Claudia. 'A Kaleidoscope Model: Defining Natural Environments', *Journal of Environmental Psychology* 16.4 (1996): pp. 335–48.

Mayer, F. Stephan and Cynthia McPherson Frantz. 'The Connectedness to Nature Scale: A Measure of Individuals' Feeling in Community with Nature', *Journal of Environmental Psychology* 24.4 (2004): pp. 503–15.

More, T. A. 'The Parks are Being Loved to Death. And Other Frauds and Deceits in Recreation Management'. *Journal of Leisure Research* 34.1 (2002): pp. 52–78.

Moustakas, Clark. *Heuristic Eesearch: Design, Methodology and Applications*. Thousand Oaks, CA: Sage Publications, 1990.

National Park Service. 'Chaco Culture National Historical Park New Mexico', https://www.nps.gov/chcu/index.htm

———. 'Lightscape /Night Sky', https://www.nps.gov/ever/learn/nature/lightscape.htm

———. 'Natural Lightscape Factsheet', http://www.concessions.nps.gov/docs/concessioner%20tools/Natural_Lightscape_Factsheet.pdf

National Parks Conservation Association. 'Destination Darkness', www.npca.org/articles/341-destination-darkness

NHS Choices. 'Do iPads and Electric Lights Disturb Sleep?' http://www.nhs.uk/news/2013/05May/Pages/Do-iPads-and-electric-lights-disturb-sleep.aspx.

Nisbet, E. K., J. M. Zelenski and S. A. Murphy. 'The Nature Relatedness Scale: Linking Individuals' Connection with Nature to Environmental Concern and Behavior', *Environment and Behavior* 41 (2009): pp. 715–40.

North Penines. 'Animating Dark Skies Project Partnership', http://www.northpennines.org.uk/Lists/DocumentLibrary/Attachments/508//RichardDarnNPennTourismSeminar2014StargazingPresentation.pdf

Ortlipp, Michelle. 'Keeping and Using Reflective Journals in the Qualitative Research Process', *The Qualitative Report* 13.4 (2008): pp. 695–705.

Owens, Steve. 'Astronomical Tourism in Dark Sky Places', www.wcmt.org.uk/sites/default/files/migrated-reports/952_1.pdf.

———. 'Dark skies on Sark', https://www.youtube.com/watch?v=_HUmuwCrbPo.

Peron, Francoise. 'The Contemporary Lure of the Is=land', *Journal of Economic and Social Geography* 95.3 (2004): pp. 326–39.

Petherick, Tom. 'Biodynamic Association, Getting Started with Biodynamic Gardening', http://www.biodynamic.org.uk/farming-amp-gardening/getting-started-with-biodynamic-gardening/.

Pike, Kenneth L. 'Etic and Emic Standpoints for the Description of Behavior', in *The Insider/Outsider Problem in the Study of Religion. A Reader*, ed. Russell T. Mc Cutcheon. London: Continuum, 1999.

Remphry, Martin. *Sark Folklore*. Sark: Gateway Publishing Ltd., 2003.

Rich, C. and T. Longcore, eds. *Ecological Consequences of Artificial Night Lighting*. Washington, DC: Island Press, 2006.

Rohde, C. L. E. and A. D. Kendle. *Human Well-being, Natural Landscapes and Wildlife in Urban Areas: A Review. English Nature Science Report No. 22*. Peterborough: English Nature, 1994.

Roszak, Theodore. *The Voice of the Earth*. New York: Simon & Schuster, 1992.

———. 'Where Psyche meets Gaia', in Roszak *et al.*, *Ecopsychology. Restoring the Earth. Healing the Mind.*

———, Mary E. Gomes and Allen D. Kanner, eds. *Ecopsychology. Restoring the Earth. Healing the Mind*. San Francisco: Sierra Club Books, 1995.

Royal Society for the Protection of Birds (RSPB). 'Connecting to Nature', http://www.rspb.org.uk/Images/connecting-with-nature_tcm9-354603.pdf.

Royle, Stephen A. *Islands. Nature and Culture*. London: Reaktion Books, 2014.

Ruggles, Clive and Nicholas Saunders eds. *Astronomies and Cultures*. Boulder, CO: University Press of Colorado, 1993.

Russell, Dale W. 'Nostalgic Tourism', *Journal of Travel and Tourism Marketing* 25.2 (2008): pp. 103–16.

Santostefano, Sebastiano. 'The Sense of Self Inside and Environments Outside: How the Two Grow Together and Become One in Healthy Psychological Development', *Psychoanalytic Dialogues* 18.4 (2008): pp. 513–35.

Scroeder, Herbert W. '*The Spiritual Aspect of Nature: A Perspective from Depth Psychology*', http://www.nrs.fs.fed.us/pubs/gtr/gtr_ne160/gtr_ne160_025.pdf.

Schultz, P. W. 'Inclusion with Nature: The Psychology of Human-Nature Relations', in *Psychology of Sustainable Development*, eds P. W. Schmuck and W. P. Schultz, (Norwell, MA: Kluwer Academic, 2002), pp. 62–78.

Seligman, Martin E. P. 'Phobias and Preparedness', *Behavioral Therapy* 2.3 (1971): pp. 307–20.

Seymour, Linda, '*English Nature Research Report Number 533, Nature and Psychological Well-being*', http://publications.naturalengland.org.uk/publication/65060.

Shaw, Robert. 'Controlling Darkness: Self, Dark and the Domestic Night', *Cultural Geographies* 22.4 (2015): pp. 585–600.

———. 'Night as Fragmenting Frontier: Understanding the Night that Remains in an Era of 24/7', *Geography Compass* 9.12 (2015): pp. 637–47.

———. 'Streetlighting in England and Wales: New Technologies and Uncertainty in the Assemblage of Streetlighting Infrastructure', *Environment and Planning* A 46.9 (2014): pp. 2228–42.

Sheehan, William. *A Passion for the Planets: Envisioning Other Worlds, from the Pleistocene to the Age of the Telescope*. New York: Springer, 2010.

Slovick, Joe. 'Toward an Appreciation of the Dark Night Sky', http://www.georgewright.org/184sovick.pdf

'Star Count 2014: A Dark Outlook for Starry Skies', http://www.cpre.org.uk/media-centre/latest-news-releases/item/3583-star-count-2014-a-dark-outlook-for-starry-skies.

Steinbach, Rebecca, Chloe Perkins, Lisa Tompson, Shane Johnson, Ben Armstrong, Judith Green, Chris Grundy, Paul Wilkinson and Phil Edwards. 'The Effect of Reduced Street Lighting on Road Casualties and Crime in England and Wales: Controlled Interrupted Time Series Analysis', *Journal of Epidemiology and Community Health* 69.11 (2015): pp. 1118–24.

Stephanides, Stephanos and Bassnett, Susan. 'Islands, Literature and Cultural Translateability', *Transtextes transcultures, Hors série* (2008): pp. 5–21, http://transtexts.revues.org/212.

Stern, Barbara. 'Historical and Personal Nostalgia in Advertising Text: The *Fin de siècle* Effect'. *Journal of Advertising* 21.4 (1992): pp.11–22.

Totton, Nick, 'The Practice of Wild Therapy', *Therapy Today* 25.5 (2014): pp. 14–17.

Tuan, Yi-Fu, *Topophilia: A Study of Environmental Perception, Attitudes and Values*. Second edition. New York: Columbia University Press, 1990.

Ulrich, Roger S., 'Aesthetic and Affective Response to Natural Environment', in I. Altman and J. Wohlwill, eds., *Behavior and the Natural Environment* (New York: Plenum Press, 1983).

———, R. F. Simons, B. D. Losito, E. Fiorito, M. Miles and M. Zelson, 'Stress Recovery During Exposure to Natural and Urban Environments', *Journal of Environmental Psychology* 11.3 (1991): pp. 201–30.

UNESCO, 'Wider Value of UNESCO to the UK', http://www.unesco.org.uk/wp-content/uploads/2015/05/Wider-Value-of-UNESCO-to-UK-2012-13-full-report.pdf

University of Edinburgh. 'Artistic odyssey to send messages to stars', http://www.ed.ac.uk/news/2016/starmessage-030216

University of Glasgow. 'Postgraduate research opportunities', http://www.gla.ac.uk/schools/ges/research/postgraduate/#/ahrccollaborativestudentship:'skiesaboveearthbelow'

Wertz, Frederick J. 'Phenomenological Research Methods Psychology: A Comparison with Grounded Theory, Discourse Analysis, Narrative Research and Intuitive Inquiry', http://www.icnap.org/wertz%20-%20paper.pdf

Williams, K. and D. Harvey. 'Transcendent Experience in Forest Environments', *Journal of Environmental Psychology* 21.3 (2001): pp. 249–60.

Williamson, R. A. *Living the Sky: The Cosmos of the American Indian*. Norman, OK: University of Oklahoma Press, 1984.

Winter, Deborah DuNann. *Ecological Psychology: Healing the Split between Planet and Self*. New York: HarperCollins College Publishers, 1996.

Wuthnow, Robert, 'Peak experiences: Some Empirical Tests', *Journal of Humanistic Psychology* 18.3 (1978): pp. 59–75.

Zhang, Jia Wei, Paul K. Piff, Ravi Iyerb, Spassena Koleva and Dacher Keltner. 'An Occasion for Unselfing: Beautiful Nature Leads to Prosociality', *Journal of Environmental Psychology* 37 (2014): pp. 61–72.

Index

C

D

St. Peter's at night © www.bournephotography.me.uk

SOPHIA CENTRE PRESS

Lightning Source UK Ltd.
Milton Keynes UK
UKHW020003150222
398672UK00006B/496